PHYSICAL SECURITY AND THE INSPECTION PROCESS

PHYSICAL SECURITY AND THE INSPECTION PROCESS

C. A. Roper

Butterworth-Heinemann
Boston • Oxford • Johannesburg • Melbourne • New Delhi • Singapore

Butterworth–Heinemann

R A member of the Reed Elsevier group

Copyright © 1997 by Carl Roper

All rights reserved.

No part of this publication may be reproduced, stored in a retrieval system, or transmitted in any form or by any means, electronic, mechanical, photocopying, recording, or otherwise, without the prior written permission of the publisher.

∞ Recognizing the importance of preserving what has been written, Butterworth–Heinemann prints its books on acid-free paper whenever possible.

Library of Congress Cataloging-in-Publication Data

Roper, C.A. (Carl A.)
    Physical security and the inspection process / C.A. Roper.
      p. cm.
    Includes bibliographical references and index.
    ISBN 0-7506-9712-1 (alk. paper)
    1. Security systems. 2. Industries—Security measures.
I. Title
HV8290.R66   1997
658.4' 73—dc20   96-43147
                                                                              CIP

British Library Cataloguing-in-Publication Data
A catalogue record for this book is available from the British Library.

The publisher offers special discounts on bulk orders of this book.
For information, please contact:
Manager of Special Sales
Butterworth–Heinemann
313 Washington Street
Newton, MA 02158-1626
Tel: 617-928-2500
Fax: 617-928-2620

For information on all security publications available, contact our World Wide Web home page at: http://www.bh.com/sec

10 9 8 7 6 5 4 3 2 1

Printed in the United States of America

Table of Contents

Introduction — vii

1. Physical Security and the Inspection Program — 1
2. The Physical Security Inspection Process — 23
3. Protective Barriers — 53
4. Protective Lighting and Utility Services — 71
5. Types of Security Areas — 79
6. Construction Standards and Requirements — 85
7. Locks, Doors, and Windows — 97

8	Facility Access Control	117
9	Security Containers and Storage Areas	129
10	Intrusion Detection Systems	147
11	Closed-Circuit Television	163
12	Automated Information Systems Security	167
13	The Security Guard Force	183
14	The Facility Self-Protection Plan	195
15	Asset Protection and Loss Control	207
16	Facility Threat, Vulnerability, and Criticality	211
	APPENDIX 1 Security Inspection Checklists	217
	APPENDIX 2 Automated Information Systems Security Checklist	241
Bibliography		281
INDEX		285

Introduction

In the total environment of security and loss prevention, the single most used method for the recognition or anticipation of any type of threat or loss to a facility has been centered on the security survey or inspection. The key factor in such programs is the ability to identify known or potential risks, threats, and vulnerabilities so that appropriate countermeasures can be developed and implemented.

Physical security has always been an extremely important facet within the security arena. Early in 1981, I was assigned to develop some security inspection procedures. Research showed that many inspection programs were being borrowed from others, usually replicated with only slight changes. Other people were then using these programs over and over, sometimes to the detriment of their organizations. I further found there was no real process or methodology in place for the development of a physical security inspection program for the inexperienced person.

There was a need to develop a comprehensive reference tool that laid out an inspection program for physical security from the initial concept through the actual inspection. Over the years I have added, revised, changed, and updated the information.

This handbook is intended to be "user friendly." The language, whenever possible, is simple in that one or more small words or phrases will be used rather than a technical term or a given level of exposition. Consideration has been given to the fact that various readers may have different educational levels and expertise in security.

This handbook provides the appropriate information within the predominant areas of physical security and loss control for conducting a physical security inspection.

The security specialist or practitioner will find this book an excellent aid for planning, determining priorities, and developing a survey program for a given facility. This book also discusses how to develop site-specific inspection criteria and checklists, how to implement the survey, and how to determine recommendations for various physical security deficiencies.

Each chapter, section, and appendix includes general and specific information concerning the subject area under discussion. The handbook addresses the requirements, needs of each security area, the applicability to different facility situations, and also the physical security inspection team members' use of the information. The appendices are directly related to each chapter and cover specific areas of concern for an inspection.

Specific chapter information will back up the various inspection checklist points and provide details to refresh each team member's memory. Additionally, the information can be used to determine baseline requirements, qualifications, specifics, and details for upgrading or creating a portion of the physical security process.

The handbook serves a dual purpose and has an increased value to the reader. It becomes a single-point reference when specific information is needed to determine security requirements within the facility; when upgrading, the handbook is useful for determining contractual requirements for work to be performed; and it is also a general reference guide for answering questions concerning facility guidelines.

Readers should be able to firmly establish and maintain an effective physical security inspection program for all sites, facilities, or installations for which they may be responsible. The book discusses approaches to program support, bearing in mind that other programs may also exist that are equally effective. The most important item to remember is that you—and only you—make the final determination as to whether or not the facility maintains the requisite level of standards for physical security . . . and this handbook will assist you to that end.

As a resource, this handbook addresses many of the problem areas you may encounter. Common sense must be your guide in areas where you do not maintain the highest level of expertise; it is probably the single most useful tool you can use to enhance the book's concepts, procedures, and guidance. It will be a valued asset to the security practitioner for years to come.

The handbook attempts to be inclusive, and thus also deals with concerns about classified and sensitive information that often pertain to the government and government contractors. From a purely business point of view, the information can translate over to "company-confidential" or "company-sensitive" material. Keeping this in mind, the reader can better appreciate the value of protecting such information, even to the extent of using government standards for the protection of corporate information.

The scope of the handbook provides physical security inspection program (PSIP) team members with instructions and examples on how to establish and then maintain an effective program to benefit the organization. It will also provide members with greater amounts of experience and knowledge, plus additional ideas to enhance their knowledge and practical experience in the proper application of the PSIP.

This handbook is not intended to change any existing policy for an organization. It is intended primarily as a reference source book and will only be of benefit and value if you are willing to accept the traditional concepts or methods of operating. Do not judge the value of the handbook solely on its individual elements; read it through and then use those portions that apply to your situation. If you see areas that would benefit from improvement, or if you have past experiences that should be included or know of a better way, then share this information with everyone concerned.

In sum, this handbook can be used effectively by private industry, government, and government contractor organizations. The basic differences are in the various required documentation and procedures that may be mandated by specific rules, regulations, and policies of a given organization. Thus, within the Department of Defense (DOD), for example, the Information Security Program applies; for government contractors, the Industrial Security Manual and the National Industrial Security Program Operating Memorandum (NISPOM) may apply; and for private industry, those standard requirements and procedures that any prudent individual or business would take to protect physical and intellectual property will apply. Each situation will be slightly different for any two organizations, even when they seem to look the same from the outside; internally, different requirements may be necessary because of what is being protected. The handbook provides a variety of topical appendices consisting of numerous questions you can use so that PSIP can be tailored to apply to each site or installation.

1
Physical Security and the Inspection Program

A definite relationship exists between the physical security program processes and the actual inspection program for any organization or facility. Each individual, whether a security specialist or another individual within an organization, must realize that good physical security benefits everyone.

Physical security is but a part of the larger overall organizational security program. It is specifically concerned with the various physical measures designed (1) to safeguard personnel; (2) to prevent the unauthorized access to equipment, a facility or installation or a portion thereof, audit materials, and information—either documents or electronic data; and (3) to safeguard all these items against espionage (corporate or international), sabotage, damage, and/or theft. Physical security, then, will include various active and passive measures.

Older programs used surveys and inspections to assess facility vulnerabilities and allowed for corrective action to increase the overall security of the inspected facility or program. Although these programs seemed to grasp the basics, they were still defensive in nature and assumed that everything must be protected. Unfortunately, this attitude meant that dollars were wasted on items that cost more to protect than they were worth. In today's economic conditions, we cannot afford to waste such dollars. This trend is expected to continue well into the twenty-first century.

Remember that, overall, you are considering PHYSICAL SECURITY. In terms of a Physical Security Inspection Program (PSIP), this includes the protective devices and measures taken to prevent a breach of the facility security; security hazards or deficiencies that can be determined; measures that can be designed and implemented to safeguard personnel from known or anticipated threats to them and to the facility; and measures that are being taken to prevent the unauthorized access to the facility, materials, and information therein.

The physical security inspection program is the bloodline; it maintains the life of your program, identifies the good, and indicates areas that need transfusion to keep the physical security process alive and well.

Underlying any inspection program is the understanding of what physical security should be doing:

1. The use of electronic security systems to reduce the vulnerability to any perceived or identifiable threat and to reduce the reliance on fixed-station security forces.
2. Integrating physical security into various organizational contingency plans, mission planning, and the testing of these physical security procedures and measures.
3. Coordination with each organizational element and/or facility area in applying physical security applications to the concepts and procedures involved in security and personnel protection and also those involved in asset protection. This should be performed in an integrated and coherent manner.
4. Training all security personnel to respond to attempted unauthorized entries at facilities and to respond to indicators and occurrences within the facility.
5. Creating and sustaining an awareness of physical security for all personnel throughout the organization via a security education, training, and motivation program.
6. Identifying those resources that will enhance and/or increase the current level of physical security at an installation or facility.

In considering these six areas of concern, inspection team personnel must be properly trained and have the appropriate physical security background so they can identify—with certainty—the various security measures in combination (such as active and passive systems, devices, and personnel) that make the system "work." Such measures will include, but are not necessarily limited to:

- Security guard forces;
- Physical barriers with active delay or denial systems;
- Locking systems, vaults, and containers;
- Intrusion detection systems;
- Protective lighting (exterior, interior, and emergency);
- Badging systems, with associated access control devices, material or asset tagging systems, and contraband detection equipment;
- Surveillance assessment systems such as closed-circuit television (CCTV); and so forth.

SECURITY SYSTEM PERFORMANCE GOALS

The overall goal of the physical security systems in place for an organization or facility is the deployment of appropriate security resources to preclude or reduce the potential for sabotage, theft, trespassing, espionage, terrorism, or other criminal activity. To achieve this seemingly lofty goal, the security system must provide the capability to detect, assess, communicate, delay, and respond to an unauthorized attempt at entry into the organization or facility environs. The various security components have functions and related measures that allow for an integrated capability, as follows:

1. *Detection,* accomplished through human, animal, or electronic means, which will alert security personnel to possible threats and attempts at unauthorized entry at or very shortly after the time of occurrence.
2. *Assessment,* via the use of video subsystems, guard patrols, or fixed-guard posts, which assist in localizing and determining the size and possible intentions of the unauthorized intrusion or activity.

3. *Communications,* through which command and control ensures that the various local countermeasures to the potential threat are all contributing in one manner or another to restrict, reduce, prevent, or contain the intrusion or activity.
4. *Delay,* through the use of active and passive security measures, including fixed and portable barriers, which will impede and delay intruders' efforts to reach their objective.
5. *Response,* by which the intruder is either deterred from the objective sought, deterred to retreat, or is captured by designated, dedicated, trained and properly equipped security forces.

Detection and delay must provide sufficient warning and protection to the installation or facility until the security response force arrives to deal with the intrusion or other activity taking place.

Each security specialist must realize that he or she can't just pick up a local facility security manual and become the site security expert. It is unlikely that anyone could adequately perform a physical security inspection without the appropriate background in the subject matter.

Physical security specialists typically have several years of experience. It may be based upon education, on-the-job training (OJT), or mentoring within the physical security arena. The specialist's experience is the very basis for which he or she has been hired. Be aware, also, that within organizations upward mobility programs exist so that the new hires at the bottom-most rung of the program ladder may only have basic formal education and not the other OJT and experience. These individuals are to be nurtured, working with experienced personnel, getting their feet wet slowly, until they understand why certain things are done in a given manner. Understanding the "why" of the process may be more important at times than just the process itself. In situations in which newly hired individuals who do not have much experience are used in the PSI process, understanding and patience must go hand-in-hand to make these individuals valued personnel in the future.

It is important to understand the casual relationships between all the factors mentioned above and then to tie them into a logical pattern for an understanding of required performance objectives, background, and training within the security arena. Once this has been achieved, it is time to address the development of a site-specific inspection program to meet the needs and desires of the organization in a manner that will best protect its assets.

Program Establishment

The establishment of the PSIP can be divided into several steps. Many readers will become associated with an organization in which the PSIP or a similar program is already established. But for those who become associated with an organization in which nothing that is remotely tied to a PSIP has been developed, the experience can be compared to having a newborn baby. In this situation, you are starting with something fresh and need to put all your energies and cares into nurturing it in a productive manner that is best for all concerned.

Let's assume that you have just been employed. You have a general security background of several years, an adequate reference library, and an abiding concern for the best possible security. Your new boss calls you in and says,

> Joe, we've got a problem. I was talking with our CEO and he is concerned about security and exactly what we are doing to protect our products, materials, and corporate information. What I need, and this comes from several discussions with my counterparts at other companies, is good physical security. Now, I know we have a physical security program here, but we don't have any inspection program to measure its adequacy and to determine whether or not it is really doing what it should be doing. I think that a topnotch inspection program is necessary and the CEO agrees. I want you to develop one. Take your time, do it right, and most of all, make it as comprehensive as possible.

The question, now, in your mind, is "Can I do it, and do it right?" Well, yes, you can!

Initial program establishment and work methodology should be based, somewhat, upon the process flow illustrated in Figure 1.1. The flowchart provides an outline of the functional activities that are performed within the physical security inspection process, the order in which these are performed, whether the actions are direct or indirect (required or discretionary), and an end indicator for future planning based on previous functional activities of the team. Remember that while developing the program, the most important tools you will be using are

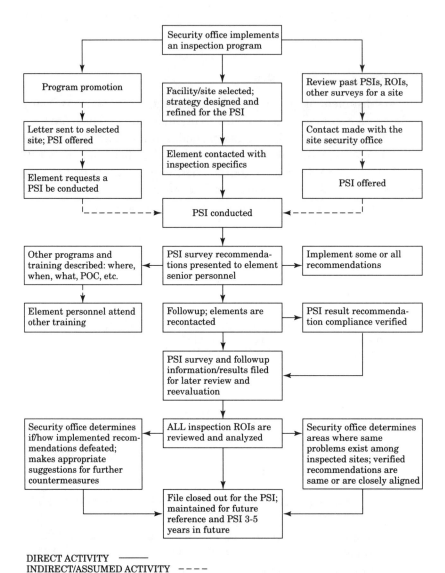

Figure 1.1 The physical security inspection framework process flowchart. Note that organizations may vary or skip some steps depending upon structure and work flow activities of the security department.

common sense and initiative. No single team member possesses all the knowledge required.

Team Organization and Duties

Chief of Security   Responsible for providing direction and oversight of the PSIP within the organization.

Chief of Physical Security/Inspection Team   Responsible for staffing, management, and day-to-day operation of the program.

PSIP Team Members   Responsible for maintaining the PSIP reference handbook and files with the latest information, being informed about the sites supported under the program, conducting physical security inspections, and advising the Chief of Security of current/projected actions that may have an impact on the program through reports and meetings.

Individuals participating in inspections should be informed about the basic functions of a physical security program; have formal training and/or on-the-job training (OJT) in physical security and the inspection process; and have specific additional knowledge of the parts that make up the physical security program. These parts are explained in chapters 3 to 16. A proper inspection cannot be conducted without the knowledge of the various parts.

The PSI team makeup is important. The team normally consists of a team chief and several members, some of whom—not necessarily all—will have unique areas of specialization.

The chief directs the program on a day-to-day and long-term basis—setting out the goals and objectives of the team, the materials required of the team, other support personnel and equipment necessary to the team's functions—and reviews inspection reports. The chief develops the overall facility inspection program and determines the inspections for each facility area.

Team personnel may be full- or part-time. The larger the facility or number of inspections performed during a calendar or fiscal year, the more likely it is that team members will be full-time. In some instances, various team members may not have a unique expertise, so additional personnel may be obtained as needed from other elements to support collateral areas to be inspected.

Based upon developed essential elements of information (EEI) about the facility that are deemed to be of a critical or survival nature, the team chief can identify the areas to be inspected and the specific qualifications required from various team members. Team member skills and assets should always be paired in relationship to the developed EEI for the facility.

PSI team members normally have five to ten years of experience in the physical security arena. This may come from government civilian or military experience or from experiences in private industry. Experiences should be fairly broad to allow for an overlapping of experience between members. Education will also play a role in team member selection. Whether such education is formal or from correspondence courses, private research, or short courses available through professional organizations, each method has much to offer and can produce excellent members. It is always good to have team personnel with a variety of educational experiences because they broaden one's mental horizons and abilities. The greater the diversity of general education and experiences in security for each team member, the better the team will be able to interact and function. The team chief should make any such requests known early on in the process to the head of physical security.

Importance of the Inspection

The PS inspection assesses the overall vulnerability of a facility to natural disasters, corporate or foreign espionage, or any disruptive actions and provides courses of action that address the threats to which a facility may be exposed. It also encourages the organization to take appropriate actions to reduce or eliminate identified vulnerabilities. The specific purpose of the inspection is to prevent loss of production and service capabilities and to provide for a rapid restoration should a disruption occur.

PROGRAM METHODOLOGY

Step 1—Learn About the Program and What It Will Accomplish for Your Organization This initial step is to gain an acceptable level of understanding about the many aspects of the PSIP. This overview is valuable whether or not you have ever performed similar duties. The education process described here is not quick or easy.

You must first develop a basic set of reference materials. It may contain general or specific publications, pamphlets, facility schematics, required areas, and/or items to be included and examined during inspections. There are a wide variety of government publications that have been developed. The government has spent many years on research and development to maximize its physical security programs, so take advantage of this and get the publications. Look back over your previous experiences and duties for ideas, publications, and references that may also apply. Search out other documents that may be of use. Other people you know in security, friends, and professional societies within the security field maintain a wealth of knowledge—much of it available for the asking. Next, read this entire handbook. Once you have familiarized yourself with the handbook and other material that may be available and pertinent, link up with other inspection team personnel and develop a list of selected facilities that you will be responsible for in terms of the inspection program.

Step 2—Acquire an Appreciation for and Knowledge of the Facilities You cannot perform an inspection of a facility without knowing everything possible about that facility relating to its physical security. You must become familiar with and understand all aspects of the selected facility operations that may relate to the PSIP. Table 1.1 contains a "laundry list" of subject area materials you should obtain for your facility reference files. A point to remember is that this information changes over time. Also, because you may be inspecting more than one facility, your reference files will expand accordingly. If a given facility has been inspected in the past, by a PSI team, by the site security manager, or by any other qualified inspection process, you should obtain a copy of the inspection report plus any other relevant supporting documentation that may apply. To acquire such information, you may need to contact the facility.

If the facility is locally based, you should have no problem. The information may also be in your files already or can be rapidly obtained by contacting the local site security manager. For facilities outside your immediate area, a formal request letter should suffice. At this juncture, please note that all documentation related to your PSI actions is important; a copy should be maintained in your local reading and reference files.

Table 1.1 A separate facility inspection file should be prepared for each facility or site to be inspected. This listing provides information about the basic documentation and facility or site description background.

Initial PSI

Location of facility
• regional and local maps
• site maps and drawings
Activity(ies) within the facility
Size of facility and physical boundaries (scaled facility map)
Number, approximate size and configuration of buildings
• individual building schematics (overall and floor-by-floor)
Type of construction used in buildings
Combustibility of facility contents
Terrain surrounding facility
Lighting for facility
Critical areas
Security levels
Size and type of work force
Utility requirements
Facility security SOP, standards, policies, instructions, etc.
Threat data based on sensitivity of facility and projects
Facility emergency disaster and reconstitution planning documents

Follow-up PSI's

Copy of initial PSI ROI
Previous inspections, surveys, etc.
New construction or major modifications since last PSI
New projects taken on by activity(ies) within facility and sensitivity
Suspected or known threat data available
Known incidents that affect facility

Based on past experiences, you may find that a local site security manager may not believe that you are "here to help." Selling the positive aspects of your inspection is a vital part of your job. Many organizations avoid the word *inspection*, preferring instead to call it an *audit*, *assistance visit*, or something similar.

When contacting the local site manager or other personnel, initiate all actions through the proper organizational chain of command. All personnel should be notified and then briefed at the appropriate time relative to the upcoming inspection. You should *never* side-step or go over a manager in the chain. This develops an enmity that will cause problems in the future.

Okay, you've developed an appreciation for the PSI program and determined the facility (or facilities) to inspect. Now, consider the development of procedures for planning your inspection.

Step 3—The Inspection Planning Process In this step, we will be referring back to some areas briefly discussed in Step 2. In doing so, we are relying on the past to evaluate the present, thus developing better protection for the future.

An understanding of the processes in the planning phase are necessary to assure a successful inspection with all physical security facets being covered. First, consider some substeps, including:

- Recognizing the need for a physical security inspection of the facility
- Stating the objectives and goals of the facility inspection
- Developing the facility background information
- Creating your plan of action
- Analyzing team members' capabilities
- Implementing the inspection plan

The establishment of this procedural sequence is invaluable prior to the actual inspection. It allows you to take the necessary steps in a logical order to develop a working model for the entire inspection. Thus, you eliminate many errors that creep into an inspection. This covers those areas that have been performed only with a stated goal, but that did not have any specific backup planning in terms of a check to be considered during the actual inspection. This ensures that you will avoid glossing over an area within the inspection and subsequent report because you have specific inspection objectives worked out before the inspection takes place and thus know fully what was done.

Recognizing an Inspection Need You have not come this far without developing an appreciation of the benefits derived from an inspection. For a specific facility, first determine if there is a "need" for an inspection. This is determined by considering the following questions and thoughts:

- Is it a new or older facility? If it is an older facility, was it under the control of another owner or was it your own facility?

This point includes any facility of an indeterminate age that has had major changes, newer risk features introduced, or the geographic area has threats that were previously unknown or not present.
- Has the facility been expanded or reduced greatly in the recent past? This includes major internal changes occurring via construction or a dramatic change in the personnel level.
- Does the current security program have known weaknesses, and, if so, are they evidenced by poor policies and procedures? Are the personnel partially or entirely untrained in all relevant areas of security? Or, have previous inspections indicated major weaknesses that, to your knowledge, have not been corrected for one reason or another?
- Do your files indicate that no inspection of any type has been conducted during the preceding five years?
- If there is a security program, is it weak? This can be demonstrated by an increased level of theft or facility damage (repeated or unreported), in addition to the loss, theft, or compromise of other equipment, materials, and/or corporate information over a short or long period of time. One should consider whether or not this activity has been significantly reduced, has stopped abruptly, or the perpetrators have been caught. But taken as a whole, this is indicative of a poor security program and a high-risk facility in terms of adequate protection.
- Have there been, or will there be, major changes in the geographical area in which the facility is located? Such changes can include an increase in population density, increased crime level, an increase in known or suspected threats to similar facilities (either within or outside the immediate area), and substantial problems, whether related or not in business, commodity, equipment, personnel, and so forth within the same geographical area.
- Does the organization have a policy indicating that inspections will be performed at regular intervals, such as once every two years or once every five years? Or is one performed only when there is a change or a specific problem is identified? In the latter case, such policy may be the basis for a comprehensive inspection. This is in addition to the reasons already enumerated for an inspection.

Using all these considerations, a determination is made for an inspection. Depending upon which factor or factors have been used to determine this need, you can now appreciate the detailed background required to fully develop your specific facility inspection.

The next portion of the planning process defines specific inspection goals.

Objectives and Goal Preparation The fundamental inspection objectives and goals are to determine the current security posture, to locate and identify vulnerabilities, and then to make recommendations—for the reduction or elimination of factors that are considered vulnerabilities, risks, or threats to the facility. Some of these recommendations may have already been determined. The next step is the development of the inspection criteria and plans that will meet the needs of the inspection. These criteria should be based on the overall goals, the anticipated behavior of the facility toward a formal physical security inspection, the time alloted to perform the inspection, and the expected results from the inspection.

A level of standards, for example, a benchmark, must be identified. Each item on the inspection checklist is measured against each of these standards. Does it meet, exceed, or fall below the standard? If the item meets or exceeds the standard, great; if it falls below the standard, how far below is it and how serious is it? Can it be brought up to standard without incurring a great deal of expense, or can the problem be remedied by changing the procedure or adding an extra step? A study of the problem usually reveals a simple answer.

In terms of results, do not focus only on the Report of Inspection (ROI), the formal paperwork report emanating from the inspection process itself, but focus on the results that are positive in terms of the overall activity—the physical security posture of the facility.

The inspection criteria and plans to meet the needs of the inspection will be developed based on the data concerning the facility. Be aware that no manager "wants" an inspection unless there are problems he or she can't fix. There should be an established schedule for inspecting all facilities, whether or not there are identifiable problems.

Facility Data Collection The collection of relevant information and specific data concerning the affected facility is of great importance to the overall successful completion of that inspection. Consider the

14 Physical Security and the Inspection Process

sources of information to be used. Official, organizationally developed sources are more reliable than others. Adequate documentation is a necessity. The types of information sought may vary with the facility location, age of facility, number of personnel on-site, facility sensitivity, and the circumstances surrounding the inspection. Table 1.1 illustrates typical information that is obtained during the facility data collection process.

Once data has been collected and sorted into general topics, a specific listing can be developed, broken down into sub-items, and further refined. The specifics to be evaluated during the actual inspection are derived from this list.

Consider any external and internal constraints that may play upon data collection. Accuracy of the information requested is a must!

- Is it time-dated or just plain "old" data that has never been updated?
- Is the data complete or have you obtained only abstracted portions of what was requested? (In other words, "what was left out?")

Consider carefully what you request and what you actually receive.

You should have a facility diagram, floor plans indicating access/egress points, badging information, a current copy of the physical security plan for the facility, statistics concerning personnel, and so forth. Review the "shopping list" (Table 1.1) and add items that are required.

Careful examination of the items can provide you with areas that may be considered for an upcoming inspection plan of action; make sure you have at least the minimum materials. In the past, individuals tended to rely upon a previously developed checklist without really examining other areas that are potentially ripe for inclusion in a proper and comprehensive security inspection.

In reviewing the data, you will make certain assumptions, including those areas that concern possible hazards and/or deficiencies that can be either manmade or environmental, and possible loss (fire, theft, natural disasters) to the facility. Such deficiencies can be based on the number of personnel working within the facility, the level of security and guard competence, any previous reports of loss or theft, the adequacy of the fire protection and/or facility disaster plans, and previous incidents that were reported but not followed up or resolved.

When reviewing the data, they should be tied, whenever possible, to the following:

- the *purpose* for which the facility exists (its mission/functions);
- the *responsibilities* of management for the facility;
- the *duties* involved for managing varied aspects within the facility; and
- the *policies and procedures* followed within the facility.

1. *Purpose.* Determine why the facility was established, exactly what the facility does, and why. What is the level and degree of competence of the security personnel and supporting security guard force? How long has the security program been in place? What positive revisions of the security posture have been performed since the last inspection? What types of files and records are maintained to back up the procedures being used? Be aware that what is visible from the exterior can affect the security measures that are in effect, either favorably or unfavorably.
2. *Responsibilities.* Who is responsible for security planning and implementation? Does this responsibility flow from upper-level management? What percentage of time is devoted to security matters? When responsibility is split, what is the chain of command? What is the authority? If the facility is located on a larger installation or within a much larger organizational structure, does authority from the installation override local facility authority, and to what extent? Is this for the good? In terms of various security features in use within the facility, does installation policy override? What concerns are there in the area of budgeting and scheduling of repairs, maintenance, and upkeep for currently installed or new devices? Are such repairs and/or device installation on a firm schedule? Responsibilities for ongoing programs, including research and development activities, must also be understood. Why does

your organization have that particular program? What is its end use: information or a product? Does the program have its own security policy and/or procedures based on a contract? What is sensitive or critical to the program in terms of information, materials, design, and so forth? Are the sensitive or critical functions directly related to the function or ability of the end product to "do its job"? Is the program known outside the facility? Do other organizations know about the program? If so, to what extent?

3. *Duties.* Where is the security emphasis being placed within the organization? Is it on personnel (on-site, permanent employees only)? On visitors (one-time access or repeats)? On contracted personnel working on the site? Is security emphasis on parking and access to the outer facility environs and then access to the buildings themselves? Is it on known or suspected threats or hazards to the facility mission and objectives? On loss prevention? Other losses from inside or outside sources? Is the emphasis on disaster planning (fire and safety of the facility, personnel, equipment, and documentation)? Or other?

4. *Policies and Procedures.* What is the basis for all policies and procedures as they relate to the facility security program? Are they outdated? Find out when were they last updated and whether or not they reflect the current functions, mission, responsibilities, and projects of the facility. A review can determine if policies have changed or been updated to meet new or sensitive programs that have been taken on during the past few years. Finally, consider if such policies accurately reflect the mission and sensitivity of the facility. What is the education and training of security and security support personnel? What general and specific manuals, references, instructions, and policies are spelled out for such personnel? How carefully are they followed? Are they circumvented? Is the security guard force in-house or hired on a contractual basis? What is the number of security force personnel on-site at any given time? Do they know their functions and duties fully? Do they carry sidearms? Does the guard force have extra duties, such as escorting visitors; locking buildings; performing building, fire, and room checks; and providing security during moves of

equipment and sensitive facility information? Are the guards properly trained?

From these concerns, you can see the importance of developing the details of the security inspection plan. It is recommended that several personnel be involved in this process to save time, especially in larger organizations. Consider all views and areas of knowledge to assure that a maximum amount of expertise is used in developing the plan. This will ensure that the inspection and the end product are as comprehensive as humanly possible.

The last two items are important in that (1) certain pieces of information need to be at the Team Chief level while others are for the individual team members, and (2) some information may be "nice to know" but may also be superfluous or company-sensitive and therefore not available except to select personnel.

Behavioral Aspects of Inspections

Psychologically, it is important to consider the reception and acceptance of the inspection team by the facility management and working employees. Acceptance of the team by employees should be to a degree that it will help to elicit the desired information from any interviews, promote the positive aspects of the inspection, and ensure success of the inspection, all without engendering ill will or negative feelings. A friendly manner and subtle persuasion on the part of inspection team personnel can accomplish more than veiled threats or an overbearing manner. Will the facility accept the inspection results and recommendations and take appropriate action to correct or upgrade deficient areas? This is where a diplomatic personality and techniques of persuasion enter into the inspection process. Remember that the inspection is not being performed to degrade the facility personnel and find fault, although this is sometimes assumed by facility personnel; rather, it is an objective examination of the security posture for that facility.

Is it likely that some information will be withheld? Certainly. If you don't specifically request it, don't expect to get it. Many of the fears can be set aside by developing a professional level of cooperation with the appropriate facility personnel. The level of cooperation means you can avoid misinterpretations or misrepresentations of inspection goals and purpose, and thus avoid limiting the efficiency of the inspection. If you don't develop this cooperation, you stand a good possibility of having an inspection that is inaccurate or incomplete and flawed because of information received either prior to or during the inspection.

Facility Security Management, Planning, and Programming

The essential elements of facility security knowledge are typically not known or are known only in a very general sense by senior management administrators because they have the responsibility for the programming, management, and continued strategic long-range planning for the facility. Security is only one management area about which they *may* be concerned, but one whose responsibilities have been delegated to the security manager. Within the security department, the security manager maintains all responsibilities for the development of organizational plans, policies, and procedures that relate to security. For larger organizations these typically include the development of the plant security organizations; the review of specific security plans and procedures developed and reduced to a written format; the provisions for reporting promptly to senior management and appropriate authorities any actual or suspected acts relating to criminal acts, economic or other types of espionage, sabotage, bomb threats, and so forth; liaison with appropriate law enforcement authorities; and responsibility for ensuring that personnel attend facility protection training courses.

The security organization must be delegated various responsibilities from senior management for the overall security well-being of the facility. The determination of these responsibilities for various phases of security and the ability to assess the performance of the persons involved are only some of many aspects that must be covered.

Liaison with law enforcement authorities is based on the determination of jurisdictional responsibility. It may be singular or multiple.

Communications within the building(s) of the facility are also a concern. It is essential to know the type and amount of communications equipment located within the facility, including trunk lines, radios, Teletype equipment, and so forth. The determination and development of security regulations for this equipment is also a functional responsibility.

Various procedures for the protection of life, property, and information must be developed and implemented. These include the key control program; a package and materials control program for the entire facility, for critical and restricted areas, and for classified products that come in and leave the facility; security and safety policies and procedures; and even procedures for fuel, oil, and other dispensers or storage of flammable material (to include warehouse long-term storage).

Vital records protection procedures for records, money retained on-site, and valuable pieces of equipment is a direct concern and is linked with the facility property and liability insurance program, if any. The nature of the facility functions determines the types and amounts of records that are maintained. This, in turn, dictates the degree of emphasis that must be placed on protection of such vital records. During the inspection, team members must view this category of security protection. If a building of which the facility is a part is occupied by other private tenants, they may have a security program already in effect for vital records protection. You may inquire if security office personnel have consulted these other tenants to determine what types of records, computer tapes, correspondence, plans, source documents, and so forth should be afforded higher levels of protection.

In terms of facility records, consider and examine the trash area. Unclassified or corporate-sensitive documentation that is discarded daily, though seemingly innocuous, can provide undesirable individuals or groups with information about the functions, plans, and future of the facility.

If the facility has any sort of medical support on-site, consider the protection of medical supplies. Strict controls over general supplies, especially any narcotics or other pain-reducing drugs, must be implemented. A survey of this area in cooperation with medical personnel will reveal the adequacy of controls that are being used to protect supplies, drugs, and equipment.

Most facilities have money stored on-site. It may be a slush fund for office parties, change from vending machines, or daily receipts from a facility cafeteria. Such areas are subject to burglary and robbery.

As you can see, security functions and responsibilities for a security staff are diverse, and some of these may not seem to be directly concerned with the facility mission. Nevertheless, anything that falls within the facility may, and usually does, have security aspects that must be considered by security and upper-facility management.

Program Monitoring

The senior assigned security person for the organization monitors the security program. At a single facility or at multiple sites, the individuals are representatives of the organizational security division.

The chief of security is responsible for administering all facets of the physical security program for all facilities and sites. He reviews all plans for construction or modification to areas to ensure that all possible physical security safeguards are considered within the design phase, that physical security deficiencies are eliminated or minimized, and that construction and/or modifications comply with application laws, regulations, and security policy, as necessary and required.

He or she also establishes security-restricted areas and determines the degree of restriction or control necessary to prevent the compromise of the various security interests. Other functions may be performed as directed, assigned, or determined on the basis of the facility mission, functions, and the responsibilities inherent to the security profession. Finally, the chief of security will provide senior organization management with current information and/or the status on all physical security matters.

Security support offices under the chief of security include personnel security, physical security, information security, and possibly, automated information systems (AIS). Within the framework of the inspection are those aspects that concern the physical security office functions and responsibilities. The chief of this office would be responsible for:

1. The design, development, and implementation of physical security policy, procedures, and measures.
2. The acquisition, installation, and maintenance of physical security equipment and other safeguards.
3. The direction and oversight of the activities of the security force personnel who are assigned to and who function for the Office of Security at all organizational facility sites and the control of personnel in terms of access, egress, property, and information at sites under the security chief's jurisdiction. At

local sites, site security managers would assume local guard force jurisdiction.
4. The coordination of contracts as they relate to physical security. These contracts would include those related to alarm systems, locking devices and security containers, closed-circuit television (CCTV), and the security guard forces at all facilities.
5. The development and implementation of a physical security program, to include all facility sites and environs, new construction, alterations or renovation to existing facilities, and special access facilities. The certification and recertification of selected facilities may or may not fall within this area, depending upon regulatory guidance.

The PSI team must be aware of the functional responsibilities of the senior security management and staff to determine the areas they are responsible for, the range of their responsibility, and the specific functional areas of responsibility that must be viewed during the physical security inspection process.

Field Program Management

Field program management applies to those facilities and elements in areas outside the main headquarters. Although management is the overall responsibility of the senior security individual or designee of the organization, day-to-day direction of the security program is vested in the site security manager.

When the security guidance and/or authority is too vague for local site security managers to take positive action, the problem should be referred to the senior individual on-site. Direct contact with this security office should be authorized, when practical, when the problem cannot be resolved at the local level.

Because inspection team members must be aware of a wide range of policies and security procedures, they will ensure that when a conflict occurs between differing instruction or local installation policies, the most stringent will be used.

For facility locations having an identified or elevated threat level, coordination *must* be effected with the local installation and/or other security offices. Local guidance will prevail when security and the threat level must be considered in the planning and operations of the facility. Site security managers should take appropriate steps to ensure the best level of security protection possible, keeping in mind that the safety of personnel is paramount. Although certain security measures and procedures may seem cumbersome, the welfare of personnel is the basis for the implementation of these measures and procedures. Team members will need to review the threat level and security protection in this light.

Team members should also look at the site security manager's individual responsibilities at each site inspected. Areas of concern include:

1. The proper administration of the security program to the betterment of the facility and the organization.
2. The development and implementation of local site security procedures.
3. The review of plans for construction or modifications to the facility.
4. The establishment of security-controlled areas according to the degree of restriction required to comply with appropriate instructions.
5. The assurance that all site elements are continually inspected and reviewed so that security safeguards are at the proper level and are maintained to provide a given protection level for the facility.

All the information that has been covered here must be considered in terms of the adequacy of the security protection provided for each facility. If a site manager becomes lax or doesn't perform certain responsibilities, there will be a decreased level of facility security. Thus, poor security is the fault of the security manager at that particular site, not the facility. Such observations are important, for if the facility manager does not perform adequately, the assumption is that security performance will probably be low also—a situation that must be corrected immediately!

2
The Physical Security Inspection Process

A LOOK AT THE STEPS INVOLVED

The inspection of a facility, or any portion thereof, is an assessment performed with a view toward *assisting* the facility security posture. The facility is analyzed as part of a whole, then recommendations are made for any improvements of the facility security level. In this regard, it must be systematic. The basic idea is to assist the local site security manager (and related personnel) to design, upgrade, and/or implement those security procedures, deterrents, and countermeasures to the affected portions of the overall physical security program. The goal is to uniformly protect the facility at a given desired level, leaving no weak links for easy attack or penetration.

It is important to be concerned with detail and to ensure continued objectivity in the inspection process. The overall objective is clear: to achieve a balanced security systems integration. For this reason, the inspection approach should be strategic. The recommendations are therefore considered strategies to stop potential incidents from happening.

In performing the physical security inspection (PSI), you must consider the prime elements of security: (1) physical security—barriers in an intruder's path, for example, obstacles that impede or are difficult to overcome immediately; (2) electronic security—devices that detect intruders before they actually succeed in reaching their goal (these include sensor devices located throughout the site); and (3) procedural security—policies, rules and regulations that are oriented toward the facility personnel.

These elements, in the proper combination, directly oppose the intruder and his or her intended purposes. Careful integration of manpower procedures and complementary physical protective aids will accomplish the best protection of the facility resources. The effectiveness of these elements dictates a clear preference for the use of aids instead of manpower. Protection aids are used to reduce manpower in various site areas, but never to the point that security would be affected by having less than an adequate number of individuals. Security aids are useful in areas that are difficult to monitor in a timely fashion. These aids supplement but never replace the human factors within the organizational environment.

The security plan for the facility must include four items that:

- detect the potential intruder at the point entry is attempted into the facility;
- deter the intruder so that if he or she does attempt to enter, the appropriate security response will be initiated at the earliest possible moment, increasing the chance that the intruder will be caught;
- delay the intruder so that he or she is apprehended before achieving the objective; and
- deny the intruder any further access to particular targets within the facility.

The use of appropriate management techniques—continual program oversight, the meshing of physical security subelements in a cohesive manner, personnel training, and ensuring that the physical security plan is comprehensive and up-to-date—will influence the role each element plays in any given situation and the materials, equipment, personnel, and information that are maintained in the facility.

Preinspection Planning

Successful completion of an inspection requires that it be well planned. Familiarize yourself with possible problem areas by utilizing office reference materials and files applicable to the specific site. At the least, accomplish the following near the start of the pre-inspection process:

1. Determine the scope of the inspection. It may not be necessary to conduct an all-out, in-depth inspection of the facility, especially if it has not had major modifications or a new mission or other responsibilities added during the past year or so; *and,* if a previous inspection was performed that had a very small number of deficiencies recorded and all were corrected or alleviated; *or* if the functions of the facility have not radically changed since the last inspection has been performed. It will depend upon the individual facility. Should the situation warrant, the entire facility may be addressed with emphasis on specific parts. This method will provide an overall picture of the operational status of the entire facility once the inspection is completed, while ensuring that the critical portion is thoroughly inspected as well.
2. Review any previous surveys, inspection reports, audit reports, and/or other pertinent data.
3. Determine the degree of technical assistance that may be required. This may include monitoring of telephone lines for security breaches, photographic support, and other specialized services not normally performed by security.
4. Develop a preliminary outline of the various inspection phases, but *remain flexible* in your outlook.

Use normal physical security guidelines *along* with common sense. Use those techniques to determine conditions within the facility that are conducive to good security practices. The inspection is not intended to provide simply a list of deficiencies, but to make the facility personnel aware of problems that could cause the facility to suffer intrusion, theft, damage, sabotage, and so forth. It also highlights positive features and shows the facility and the overall site in a positive light.

Some personnel may bring up the point of an evaluation of the management techniques used at the facility and whether or not the facility is being run economically and within current laws, regulations, policies, and procedures. This is outside your area of jurisdiction and should be referred to others.

Resurveys

When *serious* deficiencies have been identified, a resurvey must be conducted. Allow a reasonable period of time (not to exceed twelve months) for the correction of deficiencies, before another inspection is performed. A formal schedule of resurveys is not desirable but should be accomplished as needed, concurrent with other ongoing inspections, personnel constraints, and funding.

Administrative Preparations for a PSI

Once an activity is selected for a PSI and the organization, activity, program, and/or technologies are determined, the administrative preparations will begin. They will include:

1. Identifying the specific purpose and scope of the PSI. Will it encompass the entire facility or a part of it? Has there been any new construction? Are there suspected or known problems central to the inspection? Is the PSI being performed because this is a new facility or building or because a PSI has never been performed? Is it time for an updated or new PSI to be conducted because of the length of time since the last PSI, a new threat, or because something else requires that it be done?
2. Is the PSI conducted using a generally developed list of questions that could be used by anyone who wants to find out about the facility, its missions, functions, specific projects, status of new projects, inventions, research and development, and so on? Such information is considered "targeting" against the facility.

Consider developing a list of the essential elements of information (EEI) the organization should be protecting. If an EEI is not available or has never been developed, a tentative EEI listing should be prepared. The EEI will form the basis for specific operations or functions that the PSI can use to assess the physical security posture. The EEI should be based upon what the facility does, doesn't do, maintains, creates, ships, receives, and so forth. It includes both information (electronic and hard-copy), valuable items of equipment, current and future projects, any monetary instruments kept on-site, sensitive papers, locations of medical information, drugs, explosives, weapons, guard schedules, protective security features, and the like. The EEI can be developed by asking the who-where-what-when-how-why questions in regard to the mission, functions, project status, and research and development, working through each major department or division of the organization. It must be consistent with the available assets and should be clearly understood by all involved. A sample EEI subject area listing is shown in Table 2.1, below. From the list, specific EEI are developed for protection consideration.

Based upon the EEI, you can identify the areas to be inspected, the required qualifications of PSI team members, the depth required for the inspection, and other concerns.

3. Field activity facility coordination. When there are several facilities at different locations, the team chief should keep the main security office informed of all planning and any requirements. Early identification of these requirements will make it easier to arrange for whatever is required to be provided or obtained prior to the inspection.

Preinspection Schedule of Events

This should be performed by the team chief. The schedule for actions to be taken prior to the inspection is not totally firm, but it can be determined using a worst case analysis. The security personnel performing the inspection also have other duties and responsibilities within their own offices. The recommended timeframes in Table 2.2 can be modified to meet specific concerns and ongoing requirements of the team members' office.

Table 2.1 Sample listing of some of the EEI subject areas from which to develop the EEI listing for a facility. The EEI is developed based upon answers to questions that use the who-what-where-when-how-why format.

Advertising of products/services
Annual and other public reports
Anything that may provide insight into organizational situations, thinking, concerns, far-reaching abilities of company
Associations the company is a member of
Budgeting: overall; by element; personal; other
Building/site drawings/pictures
Computer programs (general; specialized)
Contracts (known; proposed)
Employees (positions; numbers; shortages of)
Engineering techniques and methods
Environmental reports
Formulas
Interviews/speeches/articles by employees/presentations and training provided to others
Legal: concerns; reports; court cases; staff specialties
Locations of all sites/facilities
Long-range goals and organizational mission
Medical
Other related organizations in field
Patents
Processes
Products
Public records checks: licenses, land use; other
Radio, newspaper, TV and magazine reports
Recruiting efforts
Research and Development
Research goals; methods to obtain goals
Resources
Safety
Security policies and procedures; security guard force; sensor locations and type
Specialized reports
Subsidiaries
Technical systems
U.S. Dept of Commerce information

PHYSICAL SECURITY INSPECTION CONSIDERATIONS

Generally, the PSI team will conduct inspections only on authorized facilities and activities. The extent of the inspection will be based on location, perceived threat level, knowledge of facility functions and mission, and the sensitivity of materials and information held at the facility.

The inspection will fall into one of several categories: (1) initial; (2) follow-up; (3) every three to five years; and (4) as directed or required due to a new situation, emergency, or major changes to the facility.

Coordination for an inspection must be accomplished before the actual inspection takes place. Use the Time-table in Table 2.2 as the minimum guide for laying out the inspection process.

1. Senior facility personnel will be informed of the impending inspection along with date(s), reason, support necessary, data required from the facility prior to the inspection, who will be performing the inspection, inspection personnel identification (when applicable). This information is provided during the 90 days preceding the actual inspection, usually in a formal letter. Security clearance data is provided when the facility is either government or government contractor. For purposes of inspection results, the information above will be broken down into specific phases of the inspection process.
2. For government contractor facilities physically placed on a government installation (including military facilities), contact the senior law enforcement office. For military installations, contact the Provost Marshal Office/Military Police (PMO/MP), and also the base engineer office for specific infomation regarding the facility. NOTE: Government contractors may have an office or entire building on a government facility. They may be responsible for internal security under the National Industrial Security Program Operating Manual (NISPOM).

Previous reports of inspections, whether performed in-house or by other agencies/activities, should be obtained and closely reviewed. These reports can shed light on previous positive aspects of the facility and also indicate areas of concern over the years.

Table 2.2 Sample inspection table of events

PSI - 90 days*. Send out letter announcing inspection to be conducted. Follow-up with specific requests for information or for coordination to have a preinspection visit to obtain necessary information.

PSI - 60 days. Coordinate the schedule, including any presurvey meeting dates. The site coordinator should ensure that personnel required, including key personnel, will be present. Ensure that all preinspection information has been obtained and is in file for review and analysis in development of the inspection plan.

PSI - 50 days. Clearance data on inspection team is passed to facility security manager. This will be necessary where government classified information is involved.

PSI - 45 days. Preinspection team meets with activity personnel to discuss objectives with local management and site security personnel. EEI reviewed and unique support identified. Based upon discussions, a formal PSI plan is made. It serves as the basis for any other support requests. Also, at this point coordination is effected in selecting specific team member makeup.

Both the coordinator and local facility personnel should begin a systematic screening of files for threat data and other information that is pertinent to the PSI.

PSI - 30 days. Team chief arranges for any flights, motel accommodations, rental cars, advance funding, etc., as necessary.

PSI - 14 days. Pass clearances and accesses for specific team personnel to facility being visited.

PSI - 1 day. Team arrives on site and begins preparations for PSI.

*Days are calendar, not work days.

Your facility file should have appropriate copies of local standard operating procedure (SOP), instructions, policy letters, security actions, and/or other guidance. Ensure that all team personnel have an opportunity to review this material for specific areas and concerns.

The PSI team general reference library should contain the latest data on items of security interest, including, but not limited to security management, intrusion detection systems (IDS), security guard force training and functional job requirements, lighting, barriers, lock and key systems, security containers, emergency destruction, facility self-protection/disaster plan, and copiers. Manuals for the use of such equipment or systems should be reviewed and checked against the actual procedures.

Finally, the security reference library must include data on any known or perceived threat to the facility or general area (including country threat data, if it is an overseas facility) including sabotage, espionage, and terrorism.

Threat data is openly available in the form of criminal statistics. Local crime reports and trends can be obtained from local police departments. Various commercial organizations provide threat and crime-related data. Although this may cost, the approach is comprehensive in scope. Also, manufacturing and business organizations may have data that can assist you in determining a crime or threat level to your organization. Where government and government contractor organizations are involved, such information may be classified. Through government security and intelligence organizations, specific studies and reports can be made available. The upper echelons of the organization's security office will be able to obtain the necessary information.

All this data is brought together to ensure that you have covered all items that must be considered (and possibly studied) during the course of the physical security inspection. If a specific item is omitted, it may mean that the inspection is incomplete, and the item may turn out to be very important.

DATA COLLECTION

Specific organizational data collection will be performed before and during the inspection. During the pre-inspection phase, it is a support function and the sooner it is begun, the better the inspection will be.

You should start developing data and specific source information as early in the pre-inspection process as possible. Ninety days is usually adequate for this purpose.

You must screen your own files, including those that date back a number of years, to obtain any information that may concern the facility. When necessary, search historical files to obtain the desired data. When feasible, and within the context of the PSI, you may also consider screening any specific known data on the facility and review and update materials from previous inspections. Extract the appropriate information and develop your own continuing database of specific facility information.

Screening *should* involve the collection of information of known or suspected efforts against the facility based on mission requirements or sensitivity of the installation. This screening would tentatively assess the potential facility threat from any possible or known adversary. An early screening allows you to develop an appreciation of how the facility may fit into a competitor's or adversary's collection efforts. This also makes it easier to tailor the inspection to the specific needs and level of security required for the facility.

Open source data can also provide a detailed picture of the security structure, its weaknesses, and the sensitivity of facility activities. For this reason, if you represent a government agency or organization, consider contacting the facility/agency public affairs office (or other appropriate office) to review Freedom of Information Act (FOIA) requests to see what type of information has been released to the media and what type of information was made available under FOIA requests from the general public. Such review should cover the past several years.

If a facility or program has a significant public visibility, consider reviewing major news publications for items of interest, including facility/activity pictures, news summaries, and the classified section of papers when specific contracts were committed to see the statements included in the advertising for specific projects or programs. Many times, a great deal of corporate or sensitive project information is indicated directly or it can be easily deduced from what is stated in job advertisements.

In viewing pictures of the facility, you may find that some of the facility security features are shown. In such cases, review the picture closely to see if it is possible to determine any deficiencies or vulnerabilities. Also, this may provide other information that you may not know about the facility from screening your files.

Team members should also examine the facility operating procedures, contracts, organizational charts, and flow diagrams. These are helpful in familiarizing team members with the operations and the scope of the activity. They also provide initial identification of the overall organizational structure, inter- and intramanagement relationships, and functional areas and activities.

All personnel should exchange ideas relating to the PSI outline for the facility. They may use sequential planning to determine who does what part of the inspection, cross-inspection and questioning of other areas, and the relationships of parts to each other and to the whole for the entire facility. This information will eventually become a topical outline and can evolve, in a short form, into the introduction to the final Report of Inspection (ROI).

Fact-Finding and Data Collection During the Inspection

Because the work so far has been preparatory to the inspection, a short discussion of the inspection process in terms of fact-finding and data collection is in order.

Because the PSI team has a limited amount of personnel and time to expend on the inspection of each facility, the team must rely on the resources and assistance of personnel at the facility to obtain further specific factual information. Capitalize on their knowledge. A key word here is cooperation. Keep in mind that the slightest inference that the end product—the ROI—may contain unfavorable information concerning a specific office (and that an individual will be identified in some way as having provided information resulting in a known deficiency) will immediately destroy the cooperative atmosphere essential to an objective, unbiased facility inspection. Thus, emphasis should be on fact-finding, not fault-finding.

If possible, and if time and resources allow, team members should also look at the physical security of the facility from an adversarial viewpoint. This gives an appreciation for what an adversary would be looking for. In this context, it may be possible to observe or determine other deficiencies that would not necessarily be noticeable from the facility point of view.

When interviews are conducted with facility personnel who are well versed in facility activity operations, you can "break the ice" with a general question, after explaining the purpose of obtaining background information on a specific project or on the facility in general. Use such questions as "if you were an adversary, how would you go about uncovering the purpose of this project?" or "how would you try to obtain corporate, classified, or unclassified but sensitive information on this project or technology?" Then follow up with "how could attempts be best frustrated from a physical security point of view?" These types of questions tend to relax the person and stimulate an informal flow of information. New insights and other areas of inquiry can also be suggested by personnel during the interview.

Your facility point of contact (POC) should arrange or coordinate the staff-level interviews with senior-level personnel; informal contact should also be made with lower- and midmanagement when interviews are not scheduled for specific times, but on an as-available basis. The POC should also make arrangements for access to various areas and provide authorization for pictures or sketches to be made, especially when the facility is located on a larger installation (local policy is always followed).

When it is a government facility that works closely with contractors, either on-site or through daily visits based on contracts, consider interviews with contract representatives. Obtain information concerning the contractors, their contracts, type of advertising they used to recruit personnel to fill the contracts, and so forth. Such information can lead to the identification of potential threat areas and may increase the facility threat and sensitivity level, thus necessitating an upgrade of the overall physical security posture.

During the inspection, team members should meet frequently, either at set times or informally over lunch or after the end of the work day. In many instances, however, the time allotted for the inspection itself will allow for only one or two meetings. During such meetings, specific findings or areas of concern that relate to other team members should be passed on so any desired action can be taken to investigate a concern, or follow up on a finding that may have other implications. Ensure that all notes of interest are as detailed and complete as possible and that everything discussed is reflected in your notes. There isn't a second chance to reinspect an area that wasn't properly inspected the first time.

If you determine that a specific finding is critical, pass it on to the appropriate facility personnel immediately so that at least temporary remedial action can be taken. Your notes should include any action taken while you were on site. (The ROI will reflect the situation and action taken in a positive manner by facility personnel.)

Because accuracy is essential, any references, document extracts, diagrams, site SOP, or other directives will have to be completely identified by the document title, date, originator, and so forth, in your notes for the final inspection report.

Facility Vulnerability Testing

There are times when vulnerability tests should be conducted at a facility. Vulnerability tests are conducted to assess the operational security alertness and posture of the security office and security guard force. They are also called penetration tests, perimeter testing, and counterforce testing measures. Their specific purpose is to actually test some portion of the facility physical security plan and the response to a security breach or infraction. Such tests must have specific objectives and limitations developed prior to the test. Paramount to all tests is the safety of personnel involved, whether or not they know it is a test.

A test could be performed to see if someone without the proper identification or authorization is able to enter a given area of the facility; whether an individual can penetrate a given perimeter line (such as a fence line) or gain access to a sensitive area without being challenged. In each, the objective is to determine if the individual is observed, stopped, if security is called. Exactly what actions are taken, or not taken, is important in determining whether or not a vulnerability exists. Tests can involve one person or a dozen people, occur during daylight or at night, be somewhat simple or exceedingly complex in the planning and operation.

Security personnel who check identification passes must detain unauthorized persons, conduct preliminary searches of suspects, enforce security procedures, and report any violations to the appropriate facility security office.

Unauthorized disclosures of information by members of the security forces or other personnel should be detected and reported immediately to the security office. This must be accomplished by those performing a test, those being tested, or by those individuals who overhear such disclosures. Even if it means halting the vulnerability test, the importance of immediate reporting cannot be overemphasized.

A means for neutralizing escorts should be considered in the interest of security, but such means do not have to actually be put into effect.

Detailed planning should be conducted prior to implementing a vulnerability test. The priority of targets, if more than one is scheduled, must be established accordingly.

Personnel who conduct the actual vulnerability test should meet all the requirements necessary and have the appropriate managerial support and equipment available. Such personnel must be in excellent physical and mental condition. Tools and/or equipment that would be considered dangerous weapons, that are illegal to possess, or that would be seen as tools used by criminal or terrorist type groups or organizations should never be used.

Any individual participating in such a vulnerability test must be briefed in all instructions pertaining to the execution of the test.

Facility senior management and the site security manager will be advised that a test is to be or is currently being conducted. If the facility security force uses weapons, then the security force supervisor (and possibly other midlevel personnel) should be forewarned. This is critical!

Procedures should be used to *simulate* the planting of sabotage devices to add realism to vulnerability tests.

Finally, a written report will provide the results of *each* vulnerability test performed. It will include recommendations (as specific as possible) to preclude such actions in the future in the case of a real threat. Distribution will be limited for such reports, and they will not become a part of the formal ROI that is presented to the facility management and security personnel.

ENTRANCE AND EXIT BRIEFINGS, AND THE FACILITY INSPECTION

The inspection itself is extremely important. You have worked for probably two or three months preparing for it—don't waste all the work through an inadvertent blunder within the first hour. The most critical part of the inspection process is right at the start—the in-briefing (see Table 2.3).

Table 2.3 Facility In-Briefing Checklist

1. Introduce yourself and team members.
2. State purpose and objectives of the inspection to management.
3. Request their views and comments.
4. Indicate (confirm) any administrative support requested/offered during your inspection of the facility.
5. In-briefing presentation:
    a. discuss thrust of inspection program as it relates to the facility.
    b. benefits of the inspection that will accrue to the facility.
    c. the potential threat (when applicable).
    d. outline how the team will operate and conduct the PSI.
6. Thank management for assistance so far, and gracefully exit, and work through your POC in the conduct of the inspection.

The Pre-inspection Briefing

Prior to the actual PSI of the facility, request a basic briefing on the facility, the activities in the facility, various ongoing projects of a government or corporate-sensitive nature, and include an overview of the direction in which the activities are moving during the next several years.

This briefing gives you a view of the current potential threat and the possibility for its increase in the coming years. This information can be valuable in considering security upgrading for all or part of the facility.

Once you have arrived on site, you make contact with your facility POC. This is usually the facility/activity security manager. The security manager should have arranged at least one hour for you to meet with senior management to provide them with a briefing of what you hope to accomplish, the direction of your PSI, and an opportunity to answer any questions.

The briefing will be an opportunity for information exchange, the meeting of key facility personnel, the introduction of the inspection team members, and determination of organizational awareness of security policy, procedures, and physical security practices in general.

At times you will find out almost immediately the concerns of the facility, and you are then provided with items of information that you might not otherwise obtain.

At minimum, team members should be prepared to discuss the thrust of the PSI program and how it relates to this specific activity; the benefit(s) to the facility activity; the potential threat (current and future); how the team will operate and conduct the PSI, how it will use any interviews, any documentation review on-site, the use of previously obtained documents, any support that was requested from outside the facility, and other data collection activities that will be performed on-site.

The entrance in-briefing is presented to senior-level facility management and security personnel. Provide them with the basics of what you plan to accomplish. With this in mind, the following is a rundown of what will be covered during the entrance interview.

1. The inspection will be a *service* to the facility.
2. The assistance and cooperation previously provided by facility management and security personnel have been extremely

helpful in allowing the team to set specific goals and objectives, and their continued assistance is further requested and appreciated.
3. Avoid the use of unusual or technical terminology.
4. Review any site waivers currently in force. These are authorized deviations from accepted standards and/or procedures. This must be done prior to the actual inspection. Concerns in this area, although they may be indicated locally as "temporary measures," may impinge upon the security of the facility. Thus, it may be necessary to highlight them in the final Report of Inspection (ROI). Clarify waivers and exceptions to policy from the outset.
5. Anticipated changes to the facility mission, functions and/or operating procedures, and the relationship of any changes to current and/or projected sensitive projects should be obtained and noted as they may affect security, threat level, or facility sensitivity. Consider any major physical changes (interior or exterior) at the facility.

The Facility Inspection

The established inspection plan should start with the inspection being conducted from the outside to the inside of the facility. Depending upon time constraints for the inspection, you may have to break down the team so that some personnel work outside, others inside. Team objectivity is maintained by having team personnel inspect different specialty areas. Whenever possible, two team members should inspect the potential problem separately, then compare notes.

In the standard inspection (from the outside to the inside of the facility), consider the facility security in these areas:

- Facility security should be observed during *all* hours of operation, from different locations and viewpoints.
- Interviews with managerial, operational, guard force, and security personnel should be performed at the earliest possible time.
- The security force should be inspected in such a manner as to not disrupt their mission in support of the facility.
- An assessment should be made of the security force training and knowledge to determine overall adequacy.

- Inspect the entry and movement control operations, but do not hinder day-to-day operations while performing this function.
- All communications used by security force (primary and alternate, base or hand-held) should be thoroughly inspected and tested.
- Each team member should cover specific areas, taking detailed notes and keeping an appropriate checklist for their functional inspection areas. As necessary, they may back up or follow other members to provide a different perspective.

The Exit Briefing

An exit briefing should be performed immediately upon completion of the inspection and analysis of data (Table 2.4). If the facility location is local to the headquarters, the exit briefing may be postponed for one or two days, but this is not recommended. The inspected facility senior personnel are anxious to see the results. When the facility is located outside the headquarters' locale, perform the briefing before leaving the facility. It is best to informally brief the local site manager first, sharing all the preliminary findings. Try to avoid any

Table 2.4 Facility Exit Briefing Checklist

1. Restate the purpose and objectives of the inspection to management.
2. Provide general information on areas of greatest concern, interviews conducted, references reviewed.
3. Provide general overview of facility security posture in terms of positive attributes, items of administrative nature corrected on spot.
4. Move into the deficiencies noted, recommendations made, and request specific comments. (Ensure all backup notes and appropriate team members are ready to support items discussed.)
5. Provide specifics on the report of inspection (ROI): when it will be ready, any future information you are still waiting for (pictures, etc.) that will be included, etc.
6. Thank management and your point of contact for support provided and assistance rendered so that inspection would go as smoothly as possible.

misunderstandings or any possible bad feelings. This ensures that there are no "surprises" when the findings are presented to senior management.

The exit briefing essentially becomes a summary overview of the ROI, which has yet to be written in its final form.

During the exit briefing, the inspection results are presented, specifics discussed, and comments made in an atmosphere of friendliness, as in an informal give-and-take discussion. The validity of the findings and their completeness in addition to any thoughts about eliminating the vulnerability are covered in as much detail as desired by management. Any misunderstandings or inaccuracies are clarified.

Let the management know that the final ROI will be furnished within ten working days, and a copy will also be provided for the site security manager. Any necessary supporting material will be included with the report.

During the exit briefing, you may wish to include threat data, a synopsis of findings, and remedial actions that have already been taken to eliminate minor deficiencies of an administrative nature—a positive feature that must not be overlooked.

Stress that the ROI is still in a draft form and may seem somewhat incomplete. In part, this is due to time limitations and constraints in putting together the exit briefing. Another reason is that the facility personnel know the facility operations in greater detail than the team members ever will.

The exit briefing, then, should provoke some thoughtful consideration of the facility security by upper-management personnel in terms of vulnerabilities and methods to reduce or eliminate those vulnerabilities.

Remember that the exit briefing, like the final ROI, is designed to provide specific information and to assist facility management in improving or implementing specifics based upon their more intimate knowledge of the facility. It should be upbeat, provide detail for significant issues only and a summary of other major items of interest, and answer any questions about specific areas that were raised by management during the initial briefing. Always point out the good points of the facility; don't just focus on what is or may be "wrong." Alienation of senior management is not a way to maintain good relations.

The following conditions must be understood by the individual presenting the exit briefing:

1. The exit interview should be conducted as soon as possible after the inspection. The inspection's goals and objectives should be restated.
2. Senior facility personnel should be informed of all deficiencies and positive attributes of the facility. By providing both, you establish an excellent *continuing* relationship and are looked upon as a "friend," not an enemy.
3. Never provide a "rating" level of the facility. This approach spells certain death for the messenger of bad tidings.
4. Recommendations should always be realistic and positive.
5. Prioritize when considering recommendations. Highlight facility features of excellence at the start.
6. A written report of the inspection will sent to the facility within 10 to 15 working days. This allows time for the reproduction of any necessary supporting documentation, and so forth.

Other items that should be highlighted during the exit briefing include:

- The inspection included a complete reconnaissance, study, and analysis of the facility and operations *as they pertain to the physical security posture!*
- A security listing assigning priorities for security resources within the facility will be developed. A copy should be provided to the facility management and security offices with the ROI, but it will also be used in future inspections.

DATA ANALYSIS

The PSI team chief is responsible for ensuring that all findings will be based on an adequate collection of data, and that any findings are supported by specific data. Because of this, objective analysis of data is extremely important.

Analysis of data is the final step in the development of the ROI. It starts with a developmental listing of the deficiencies noted by team members. The team, as a group, reviews all deficiencies (indicating specific vulnerabilities and unique site characteristics), determines which vulnerabilities relate to others, and may also indicate other concerns that would not have been found otherwise.

Don't try to develop any specific analogies or draw inferences that were not actually observed or considered during the inspection process. This leads to inaccuracy and can affect the overall objectives of the inspection itself.

The results of the data analysis should be organized in a format and line-item-type inclusion as it relates to the final ROI. It is usually more convenient to group the information into functional categories, related to specific areas of the site physical security layout or to specific subject areas based on the inspection checklist.

By maintaining the data in a systematic format, it becomes more manageable and makes referencing more convenient. Also, this allows for the cross-checking of information from team members' notes.

Measure each finding against possible facility threats. For the present, consider such findings without the amount of detail that you will be using later in reviewing the facility security. (This is an important reason to retain all notes from the inspection until the postaction review is completed.) A review of a threat against a specific finding may show that it is not really a vulnerability, but more of a site hazard or administrative problem that can be easily remedied. Again, make sure that your findings are specific and supported by detailed notes or other information.

In other areas of concern, the findings may appear to be aligned with a facility threat, but the relationship to a sensitive area or project may not be apparent to team members. Consider whether or not an adversary could develop an indicator and make it useful.

The ultimate goal of the PSI is to ensure that the facility security is at a level that any such indicators and vulnerabilities are not obtainable by adversary personnel. Such information may not go into your ROI, but can be used as general follow-up information to the facility indicating specific concerns that were noted during the inspection process. Such information, while of an operations security (OPSEC) nature, can prove invaluable to the facility management, and it provides you with some extra "points" for future use in terms of support and liaison with the facility personnel.

THE REPORT OF INSPECTION

The report of inspection (ROI) becomes the culmination of the PSI team effort at the facility. It provides a detailed picture of what was observed, inspected, and considered and also lets facility management know where your concerns were emphasized.

During the course of an inspection, the inspection team members will be observing conditions, problems, and/or deficiencies that must be included within the ROI. Whether these observations are positive or negative in terms of the facility physical security, the discussion within the ROI must give them equal weight, for the ROI is a summary of your actions and the activities of the facility and its personnel, thus team objectivity in all phases is mandatory.

This section discusses concerns and provides comments and points of view upon which to base the determination of deficiencies and any proposed recommendations that would be incorporated within the ROI.

An important point to remember throughout the inspection, which is especially important in terms of the ROI, is that each team member has certain responsibilities. Among these responsibilities is the requirement to be objective, to report all findings, whether positive or negative, and to develop *realistic* recommendations based upon deficiencies that have been found. This may mean that in addition to using the facility checklist, you will have to take specific notes, interview personnel, make diagrams, and take pictures, in addition to checking that local security references are on file and up to date. This detail will support the deficiencies, recommendations, and conclusions that are a major part of the ROI.

The ROI follows a format, a sample of which is provided at the end of this chapter (Table 2.5). The ROI may be drafted as you develop the various portions of the inspection, in terms of format or placement of items, but it should be left as a draft. Upon completion of the inspection, you can collect the various items of information, develop specific deficiency statements, organize them in a logical order, and follow each with the appropriate recommendation. At this time, make necessary editorial and other changes to the ROI draft before preparing the final copy.

Please note that the ROI sample format is not "firm," but includes all items that you might ever need in a report. In instances in which

Table 2.5 Sample Report of Inspection

Office Code   Date

SUBJECT: Report of Physical Security Inspection of (*Facility*)

TO:

1. Introduction
    a. A physical security inspection of the (*FACILITY*) was conducted during the period (*DATES*) (*in response to a request/as directed by*).
    b. The following personnel conducted the inspection:
    Adams, John B.
    Jones, Alfred N.
    Perawinkler, Rhonda M.
    c. Mr. Josh Charles, Jr., facility security manager, accompanied team members and provided local administrative support and work space in support of the inspection team. His assistance is greatly appreciated.
    d. No record of an existing PSI report was found for this facility, although previous inspection reports, a counterintelligence survey, and a local site security survey were located. All are over 3 years old.
2. Facility Background. A summary of the facility, describing in general the size, area, number of buildings, sensitive areas, security concerns and responsibilities, security guard force, etc.
    a. The (*FACILITY*) is a part of the (*ORGANIZATIONAL NAME OR ELEMENT THEREOF*), supporting (*NUMBER*) elements (*ANY FURTHER IDENTIFICATION*). A total of (*NUMBER*) work on site, with approximately (*NUMBER*) of visitors during a 30-day period.
    b. Based on a review of the facility security posture as found during this inspection and through personal observation by team members, the level and degree of security required for this facility in order to protect essential mission functions and responsibilities is (*LOW, MEDIUM, OR HIGH*).
    c. The mission of the facility for (ELEMENTS A & B) is Unclassified. The mission is continuous, because the organization has continuing requirement for the technical expertise and computer system specialities to meet overall mission objectives as directed by its charter and as further directed by higher authority within the boundary and framework of the charter.
    d. Functional areas of the facility elements include:
        (1) A computer facility processing information up to and including (*SENSITIVE LEVEL*) information. Also, other special information is processed in a computer facility subsection that maintains a separate security system that is also tied in to the central monitoring panel.
        (2) A major research and technical library, to include books, magazines, microfiche, and an on-line computer system for tying in with other agencies of the U.S. government for obtaining on-line computer assistance and reference documents.
        (3) Research and development of computer-enhanced communication systems, modeling, techniques, and approaches to communications in the out-years.
        (4) Major coordination efforts exist between both elements and three of the contractor facilities in support of a major mission requirement.
3. Survey Findings.
    a. Overall, facility maintains a security adequacy above that acceptable of the activity being inspected and must be convinced that the team report will not be the basis for disciplinary action, but one of positive security reinforcement for the facility as a whole. Concerns of this type can destroy the team's ability to collect factual and unbiased information.
    b. Should a significant security weakness be found, immediate follow-up is neces-

sary. It must be accomplished in a constructive manner without assessing individual blame or responsibility.
- c. Each PSI undertaken is unique in itself, and it is objective in its accomplishment. The ROI, as the end product of that PSI, must also be objective and free from fault!
- d. The ROI has several major headings, and then subheadings, which can be used by team members to develop their individual portions. The major portions include:
    (1) Introduction: When, where, why a PSI is being conducted. Who the team members involved are. Who provided support and assistance at the facility level. Any record of previous inspections and comments on frequency of past inspections or reports by a PSI team or other command/agency.
    (2) Facility Background. A short summary indicating what facility was inspected, various elements located within the facility, a general determination of the facility level, and the functional areas of the facility elements.
    (3) Survey Findings. This is by far the largest portion of the ROI, including both positive and negative determinations for the facility, and the appropriate recommendation for each deficiency found. Further, any deficiency that was corrected on the spot, prior to the teams completing the inspection, is included with a notation that the deficiency has been corrected. Sample deficiency write-up could be as follows:
4. Survey Results. (AT THIS POINT A SUMMARY OF DEFICIENCIES WOULD BE INSERTED, IF APPROPRIATE; SUCH A SUMMARY WOULD BE LABELED AS PARAGRAPH THREE).
The following deficiencies were noted:

DEFICIENCY #1 - Security Guard Post, Main Entrance. Broken windows on side and rear have been covered over with cardboard. Additionally, the light above the guard post is of a reduced wattage (75 watts) and is insufficient to determine proper identification of vehicles and personnel entering the main gate or pedestrian gate located directly behind the guard post.
RECOMMENDATION. That the broken windows be replaced, that a better illumination system, preferably small spots, be installed to allow viewing on all sides of the guard post and pedestrian walkway.
NOTE: Glass has been replaced, and several lights are currently being installed to eliminate deficiency.

DEFICIENCY #2 - etc. . . .
5. Conclusions.
- a. The (FACILITY) has a better than average level of security awareness on the part of building occupants.
- b. The physical plan for the facility, the self-protection, the disaster control and facility reconstitution plan, and the local implementation of security policies are effective, correct, and positive. There is no problem with compliance on the part of the facility occupants.
- c. Of the twenty-eight deficiencies found in nine areas, seven were resolved during the course of the inspection timeframe.
- d. A high standard of security awareness by security guard force in the implementation of security policy and procedures must be recognized. The security guard force is a commercially contracted force and has standards and professionalism equivalent to a facility-sponsored security guard force.

7 Enclosures
a/s

JOHN B. DOE
Facility Inspection Team Chief

specific item areas are not included, you can determine where they would logically apply based on the ROI format.

The ROI will specify from the start the purpose, scope, and constraints that may have influenced the final report. It should include the inspection methodology (mandatory), EEI (optional), threat data (optional), acknowledgements of facility support (mandatory), explanation of areas not included in the ROI (mandatory), and an overall summary of the facility (mandatory). The facility assessment should be in sufficient detail that management can understand clearly the specific findings and reasons for the recommendations. When threat data is included, it should be linked to specific EEI and specific findings of the ROI.

Because of the information contained within the report, it must be treated in every respect as company-sensitive. Thus, it needs to be protected at all times. To see this in a different light, imagine what could happen if the information were obtained by a competitor, a foreign country seeking information about a product or service or the latest research and development product results, or if a criminal element learned of vulnerabilities affecting the business. When an ROI is for a government entity, appropriate government security classification rules and protection measures apply. In either case, keep all copies to a minimum.

The development of the ROI requires that a standard format layout be used by all teams for consistency. Table 2.6 provides a subject outline of the minimum amount of information that is included in the report of inspection.

The following summarizes the various ROI portions:

1. The opening paragraphs cover the purpose of the inspection, the facility inspected, the location, and the personnel conducting the PSI.
2. Next is a brief examination of the facility activity, to include pertinent history of the facility and the building, the mission of the facility, length of time in operation, and the sensitivity of the facility. At this point, include significant EEI and threat data, if appropriate to the facility and the ROI.
3. The next paragraph discusses the methodology and techniques used during the inspection, including interviews, vul-

nerability testing, areas of special concern or interest, areas that receive only minor inspection, and a discussion of references necessary to the inspection.

4. The major portion of the report is next and includes the most critical sections—the positives and negatives of the inspection.
   - Begin with the areas that had an adequate or more than adequate level of security. Start with the best features of the facility, specifically commenting on the most positive aspects of security.
   - Follow this with the deficiencies and specific recommendations to alleviate such deficiencies. If not addressed previously, include items that were identified and corrective action that was taken care of on the spot.
5. The final portion of the ROI will be a "thank you" for assistance, any comments on follow-up spot-checks that may be performed (in person or by letter), and expectations for the future.

Table 2.6 Outline of the format for the Report of Inspection

1. Purpose of inspection. Include facility identification, location, and personnel conducting the PSI.
2. Facility background. Summary history of facility and buildings, mission of facility, length of time in operation, and sensitivity of the facility.
3. Inspection parameters. Limitations of the inspection, methodology and techniques used, vulnerability tests conducted, special area concerns, references used.
4. Positive aspects noted during the inspection.
5. Deficiencies and recommendations. Indicate specific deficiencies, tying them to the EEI and threat, if applicable, and support them with realistic recommendations and background, if appropriate. Specific supporting information such as pictures, sketches, facility documentation extracts should go as enclosures, but are referenced herein.
6. Summary.
   a. Summary of inspection.
   b. Assistance rendered (a "thank you").
   c. Follow-up/coordination in future.

Enclosures: as applicable

All in all, the ROI will be developed in a positive vein that demonstrates objectivity.

In developing the recommendations corresponding to each finding within the ROI, keep in mind that some may not be easily implemented at the facility level, others will be simple to implement, and still others may involve a substantial cost that has not been budgeted.

It becomes the responsibility of the facility management personnel to determine which, if any, of the recommendations will be implemented. The ROI then becomes *their* management tool to allow the facility management to make decisions affecting *their* security posture. In other words, the ROI recommendations are just that—recommendations for consideration by senior management.

The completed ROI should be a professional piece of work. It must be error free and letter perfect, covering all areas with the same level of consistency. It should never be slanted to conform to a particular viewpoint or to soothe feelings, but it should provide specific and accurate information regarding the current physical security status of the inspected facility.

You may think that local politics influence the inspection and the ROI because the ROI may be seen by others than just certain individuals of the inspected facility. This should never be the case. The ROI is an important document that must stand by itself. In this regard, remember that politics can change with administrations. By staying objective, you provide information and neither support nor deny a specific local viewpoint regarding the facility preparedness and its security level.

The inspection is a fact-finding process. Your function is not to examine compliance with specific security directives, but to assess the actual security of the facility in terms of the appropriate security directives and common sense factors that dictate all normal actions. You must avoid appearing as "fault-finders" or "investigators."

The conclusion portion of the ROI is a summary of the facility security level; the positive attributes and deficiencies/recommendations are again mentioned, but not in detail. A short statement concerning facility support, in terms of personnel and logistics may be included if appropriate. Specific individuals may be mentioned by name when they have provided an exceptional amount of support assistance.

Sometimes, the ROI conclusions will be specific in detail, while other times the specifics will be found in the conclusions and the follow-up details will be elsewhere in the ROI. Finally, the ROI should be signed by the inspection team chief.

Identifying Deficiencies and Making Recommendations

Develop all deficiencies fully, stating them in a clear and concise manner. The inspecting team members should not equivocate. It is extremely important that all statements relating to a deficiency and the associated recommendation be presented objectively. Personal observation of a potential or known deficiency enhances the team member's insight and makes the determination of the recommendation specifics more acceptable. There must be a sound basis for making the recommendation (Table 2.7).

The ROI recommendations derived from the information developed will be useful only when presented in a clear and concise manner to the facility recipients. All such recommendations must be based on a specific problem that was noted, not upon supposition or assumption. The recommendations must be realistic and feasible, and the interpretation of them must be "crystal clear." Recommendations must:

1. Allow the recipient to be positively motivated to act. In other words, the recommendations must be realistic and feasible.
2. Establish a frame of reference (clearly stated problem).

Table 2.7 In stating a deficiency, the writer needs to be as specific as possible and the recommendation here states how to eliminate the deficiency. An inspection team note may indicate a past problem or concern that has security implications.

DEFICIENCY #3 - Building #4. The rear (south side) first floor window toward its left side has broken panes. Panes are broken *inward;* evidence of broken glass still remains on inside floor. Open area of window is an area of 18" × 27" and can allow an individual to enter undetected while building is not being used by individuals.

RECOMMENDATION: Window should be sealed entirely, or a protective steel mesh screen used to protect interior from unauthorized entry.

INSPECTION TEAM NOTE: Building has been broken into twice during past year, and only the window glass was replaced. Suspects in both break-ins have never been apprehended. Office equipment was taken in both thefts.

---

3. Demonstrate a realistic expectation (enhanced or reduced security level that is acceptable to site personnel).
4. Alter or reinforce any preconceived ideas regarding the threat level or potential risk to the facility (feasible and acceptable alternative to poor or reduced security practice or standard).

The deficiency and recommendation for each is most important; it must be factual, concise, and detailed (Table 2.8).

A photograph or sketch of the area in question can be included as an attachment to the ROI.

Sometimes the list of deficiencies is lengthy. Summarize the deficiencies in a general fashion and insert them near the start of the ROI following the first paragraph. General comments can apply to various areas having the same types of deficiencies.

THE INSPECTION CHECKLIST

There is no specific, prescribed inspection checklist. However, there will be variations based on different missions and functions of inspected facilities, even those within the same overall organization. A checklist of a general area would not be appropriate for inspecting an oil field refinery and a computer facility, for example. Each has different functions and areas of importance. For that matter, each checklist is different in its emphasis for each facility.

Different levels of security awareness and physical security posture may be required, depending on a facility's mission, location, factors

---

Table 2.8 Although somewhat short, the statement is concise. It indicates deficiencies and also some on-the-spot corrective action that was taken.

XX. Perimeter security. Overall, the perimeter security for the facility is above acceptable standards and levels, but three deficiencies (items 3, 11, and 28) were noted that can be resolved at a very minimal or no cost except for on-site labor that should not exceed four hours total. In this regard, item 3 was corrected on the spot by changing guard patrol procedures to include the checking of certain building areas not heretofore checked on a routine basis.

contributing to the current security posture, threat level, and sensitivity of mission-related or directed activities. This may also vary at different points throughout a given facility or at different facilities of like activities within a given organization.

The appendices provide checklists that can be used for inspections. These listings are categorized by subject areas. Either use the entire subject listing, avoiding the items that are not applicable at a particular facility, or else extract those items of concern and develop your own site-specific checklist. With minor deletions and additions, you will have a series of subject checklists that can be used for a specific site and referred to again whenever that site is scheduled for an inspection. The checklist(s) become more useful over time as they are refined in conjunction with changes to the facility mission and functions.

Make copies of the final checklist and put a copy in the facility permanent reference file. Another copy can be given to the facility security officer for information and future reference for performing self-inspections as changes occur. A copy should go to team members, with their specialty area(s) being the uppermost portion, because that is where they will concentrate their efforts during an inspection.

Upon completion of the inspection, team members can further refine the checklists of specific areas.

Because it is difficult, if not impossible, to design a single, universal checklist to cover all applications, checklists for different types of facilities will vary in length. The designer of the initial checklist must use all the information available from previous inspections to compile the best checklist for a given facility.

The length of the security checklist depends upon the facility site location, various local factors, and the amount and level of security that will be required to maintain a given level of protection. Some sites the author has visited maintain a guard on post for visitor control and access purposes only during weekday working hours, based on a time-phased entry/exit system (twelve hours a day); other sites maintain a twenty-four-hour-a-day guard, seven days a week. From your own background you will find other examples that will affect your checklist of what to include and what to exclude. The determination is up to you.

3

Protective Barriers

Barriers are used (1) to define the specific physical outline of the facility perimeter, various buildings, and access areas and points; (2) to deter illegal, unauthorized entry to the facility grounds and internal environs; (3) to delay intrusion into a controlled area; (4) to economize the use of the security guard force; and (5) to control and direct the flow of traffic, either pedestrian or vehicular. Throughout this chapter, we will examine the various types of protective barriers that can be used and discuss how they enhance the physical security process.

Barriers fall into two major categories: natural and structural. Natural barriers may be streams, desert, hills, rivers, and so forth. Structural barriers, which we will be concerned with here, include such manmade devices as fences, walls, floors, roofs, grills, bars, roadblocks, and/or other construction that makes penetration difficult.

From a security protection viewpoint, barriers have two aspects. First they create a psychological deterrent, and second, they have a direct impact on the number of security guard posts required and the frequency of use for each post.

In considering these two aspects, bear in mind that while the barriers will be defining physical aspects of the facility, it is hoped that they also will optimize the use of security force personnel while enhancing the capabilities for detection and apprehension of unauthorized individuals.

BARRIER CONCERN ANALYSIS

Facility Security Planning

The goal of good security planning is that in-depth security takes into consideration the mission and function, environmental concerns, threats, and the local area outside the facility confines. This can be translated into an A–B–C–D method. Using this method, you can have an excellent security protective level.

- A. AIDS to security. The use of sensors (area protection in an exterior mode or a large room, point protection for a room door or on a security container; and general protection as required).
- B. BARRIERS for security. These can be buildings, fences, temporary check points, and so on. Figure 3.1 shows an exterior fence, an interior fence around a controlled-access building, the building itself, and a room within the building, in addition to the security container within the room. Each of these is a barrier.
- C. CONTROLS. Controls support the physical security barriers, such as an access control system, various level identification badges and temporary badges, security escorts, and internal procedures.
- D. DETERRENTS. Guards, lighting, and checkpoint control procedures are a few of the deterrents that ensure that it will be difficult for an intruder to successfully gain access. Figure 3.1 shows gate guards, building entrance guard, a lock on door to room, and the security container requires knowledge of the combination, in addition to a room alarm sensor that is monitored from the main guard post.

When properly used, the A–B–C–D aspects double within themselves to support each other. Thus, a control is also a deterrent, and a barrier, if need be. By combining the A–B–C–D, sufficient obstacles are created to prevent an intruder from getting the information that is being worked on during the day in the controlled-access building and then is protected at night, weekends, and holidays from general theft through implementation of the security in-depth concept.

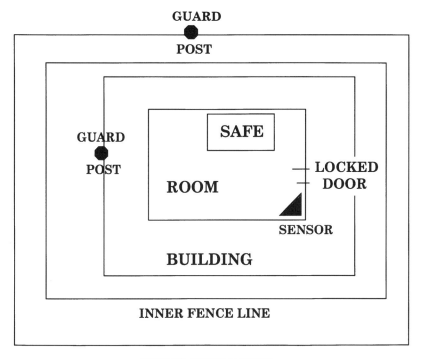

Figure 3.1 For access, an individual must be authorized to pass through a series of barriers (including specific checkpoints) to gain access to the information contained in the safe. The authorized person would have been issued a room key and been authorized to have the safe combination.

Barrier Considerations

Barriers should be used to protect the entire facility and to establish any restricted or controlled-access areas. Define the outer perimeter of a controlled area, for example, by using:

1. Structural barriers at control points of possible entrance and exit.
2. Barriers between control points that are sufficiently obstructive and difficult to traverse—to control and preclude accidental intrusion.

Positive protective barriers should be established for (1) controlling vehicular and pedestrian traffic flow; (2) checking identification of personnel entering or departing the area; and (3) defining a buffer zone for more highly restrictive areas of the facility. All barriers should be designed in view of the threat, deterring access to the maximum extent possible.

Next, positive barriers are required for the entire perimeter of a controlled, limited, or exclusion area. They cannot be predesigned for all situations; thus, they should incorporate the following elements:

1. Structural perimeter barriers, such as fences, walls, and so forth.
2. Provide restrictions at points of entrance and exit for identification checks by either pass and badge exchange, badge examination, or electronic access system.
3. Opaque barriers to preclude visual compromise by unauthorized personnel could be necessary in some extreme instances.

Always keep in mind that some barriers may delay but rarely stop the really determined intruder. To be effective, barriers must be augmented by security forces or other means of protection and assessment of potential intruders or other unauthorized personnel at the earliest possible moment.

Barrier Determination

The type of barrier to be used at any given point or area for a facility is determined only after performing a study of local conditions. During the inspection process, see if the barrier was there prior to the activity's moving into the facility. Was it constructed specifically because of the activity or after the activity moved into the facility? Depending upon the answer, you may want to take a closer look at the barrier. A barrier, or combination of barriers, must afford an equal degree of continuous protection along the entire area the barrier is designed to protect.

In cases where there is a high degree of relative criticality and potential or actual vulnerability to facility operations, it may be necessary—and is most desirable—to establish two lines of perimeter barriers. Such barriers should be separated by not less than twenty to thirty feet for optimal protection and control.

Such perimeter boundaries should be fenced and posted to establish a legal boundary. This defines the outer perimeter, provides for a buffer zone to be established, facilitates control of entrance and exit, and also makes accidental intrusion extremely remote and unlikely.

In developing these outer barriers, consideration must be given to the provision for emergency entrances and exits. Such openings, as any other for the facility, should be kept to a minimum. This is consistent with the efficient and safe operation of the facility without degrading the established minimum security standards. During an inspection, it will sometimes be difficult to establish whether or not this has been taken into consideration in the planning process. Discussions with the facility security manager and a review of emergency disaster/recovery plans and procedures will greatly assist.

During the inspection process, you should also examine recent, current, or planned construction that has a barrier application, as well as the removal of any current barriers. Has the work been reviewed and approved (in terms of security protection) by the security office? Will these changes ensure that the addition or removal of a barrier continues to provide a proper level of security?

These are typical questions every security manager should be asking. Table 3.1 provides a list of considerations for a barrier determination regarding location and use.

BARRIER REQUIREMENTS

Fences

The most commonly used perimeter structural barrier is the chain-link fence. With some modifications, the standard fence can be upgraded to provide a higher level of security protection against intruders. In the majority of cases, intruders go through or under a fence instead of over it. The simple reason is that there is less chance of initial discovery. Guard personnel can more easily see an individual climbing over a fence that is at least seven feet high (this includes any outrigger on top of the fence) than an individual who cuts out a small portion of a fence and crawls through, or the person who crawls under the fence.

The chain-link fence should have a minimum height of seven feet, although eight feet is becoming more common for most facilities. The fence should be of 9- (or at least 11-) gauge steel wire chain with mesh openings not greater than two inches square, and it should have twisted selvage at the top and bottom.

Table 3.1 Preparing for barrier determination in terms of location and use to the facility

Preparing for Barrier Determination in Terms of Location and Their Use

1. Physical layout of the facility area. Review maps (current, older maps, and projected designs), sketches, and photographs showing the area, types and location of buildings, roads, sidings and piers.
2. Consider and review roads and highways passing or leading to the facility. Determine what access routes are in the area and how close to the facility they lie.
3. Consider the physical description of the surrounding area, to include the various topographical features, adjoining property, in terms of terrain, buildings, etc.
4. Look at the total facility ground area, the location of current buildings, and anticipated placement of future construction. What is the distance from fence line, entranceways, line of sight from other nearby buildings into windows of your facility buildings?
5. What type of initial perimeter barrier does your facility use? Is it consistent with the known or projected threat level against the facility? Does it take into account the known local crime problem?
6. Look at the support-type buildings on the facilities. Do all the buildings really support the mission and functions of the organization? Are loading docks, platforms, sheds, storage, engineering, heating, power, electrical boxes, telephone controls and switching boxes, motor pool, fuel storage, trash storage, communications systems and wires, and transformer stations provided with an adequate protective level? Do they require specific barrier controls around any of them? Will such barriers cause only inconvenience, or major problems? What about local codes and laws; will they affect placement of barriers and access/exit points and controls?

The fencing *must* extend within two inches of the ground when ground is firm and hard, or below the surface at least four to eight inches when the soil is sandy or is easily windblown or shifted. Cover the soil with gravel if it can be windblown or shifted.

The height does not include a top guard. Nor does the standard height take into consideration points at which the fence is connected to a building. At such points the building connections should be higher; twelve feet is the recommended height at these connection points.

The fence meshing should be taut and securely fastened to rigid metal fence posts. All fence fabric ties should be of the same gauge as the fence. To ensure the rigidity/tautness of the fence mesh, additional support bracing should be considered as needed at corners, gate openings, and areas along the fence line where natural elements can interfere with the fence.

Culverts and similar openings along the fence line, either those passing through or under the fence, should be of pipe not greater than ten inches or in clusters of equivalent diameter. Where openings are greater, consider using grates or grills affixed over the opening. They can be permanent or removable, with removal controlled by locking devices.

All perimeter fences should have a top guard that is angled outward at a forty-five degree angle, pointed upward, and has three to four strands of barbed wire stretched tautly. This top guard extends the vertical height of the fence by at least one foot. The top guard supporting arms must be permanently affixed to the top of the fence posts. The barbs in the wire should be six to eight inches from each other. For that portion of the fence immediately adjacent to a gate or other opening, the extended height should be eighteen inches and is usually vertical to allow for the opening of the gate.

The purpose of a fence can be defeated if the fence is located so that the topography allows an intruder passage over, under, or around very easily. In such cases, the fence will have to be modified or extended or a different type of barrier structure will have to be used.

There are enhancements that can be made to fences to increase the security posture. While they may look good, they are expensive, and the return from such enhancements is not readily seen. Intruders will consider the enhancements and look for another method of entry, usually the one they had planned in the beginning. Remember, for the intruder, it is easier to avoid being seen by entering from the bottom instead of the top of a fence.

Enhancements can include "draping" of barbed wire along the top of the fence to the inward side. This uses up a lot of wire and serves no great purpose unless a large quantity is used all along the fence line.

Barbed wire concertina, such as that used in military field applications, can be used to provide additional barrier protection at the top of the fence. In addition to the difficulty of climbing over it, it will also increase the height of the fence by two to four feet.

Some fences include the military grade General Purpose Barbed Tape Obstacle (GPBTO) on top. The GPBTO is extremely sharp, having razorlike sections that easily cut into the hands and arms and can cut right through clothing. Like concertina wire, it creates a very difficult point of penetration, can be dangerous to the individual who decides to climb over (such as the teenager on a "dare"), and may be prohibited in some localities.

If GPBTO is ever used, consider local laws. Check with the legal office to ensure that GPBTO is always *above* the ground—mounted—never on or near the ground. Even within your property line, when GPBTO is placed on the ground, you are "inviting" potential lawsuits.

During the past decade or so, the increase in security awareness has been tested. Results of tests are available from various sources. In terms of fences, tests had penetration attempts on the various types of fencing. Penetrations were conducted under, over, and through various configurations. The results indicated that determined (namely, professional) intruders preferred to go under or through, rather than over fences because of the exposure factor. Seven-foot chain-link fences, with or without a barbed wire outrigger, provided less than fifteen seconds climbing delay to the intruder. A man-sized hole can be cut in a fence in less than twenty-five seconds. Fencing anchored in firm ground or in concrete will prevent a rapid penetration under the fence fabric using a pry bar or similar implement. The most feasible configuration tested was a fence using a double outrigger. A roll of GPBTO mounted on top of the fence, while a most formidable obstacle, provided only minutes of climbing delay to the most serious intruder when the intruder has been properly prepared to defeat the obstacle. This type of configuration was also the least costly and provided the most reasonable delay to unauthorized entry. Also of note is that the GPBTO was affixed to the top of the fence, thereby reducing the hazard potential to children and animals.

These barrier tests viewed such penetration and the associated perpetrators as the most determined class of professionals. Individuals not at this level would take much longer or would be effectively deterred.

Walls

Walls, when used as barriers, have some advantages and disadvantages, which are shown below at Table 3.2. Whenever possible, however, consider the use of standard walls versus chain-link fences.

When situated close to a building, a wall provides a much easier route for intruders. The downside, whether the wall is close to a building or not, is that the security force personnel cannot see the other side, so a wall can be used to the intruder's advantage to avoid quick discovery.

Walls should be of masonry block or brick. They should have a minimum height of seven feet and be topped with a barbed wire guard, angled outward and upward at forty-five degrees. If there is no top guard, the height should be eight feet and the topping should be a thin layer of cement embedded with broken glass fragments of varying sizes. The wall should provide protection greater than or equal to any other barrier.

A long-term problem will arise because of blowing sand and dirt over a period of time. It will wear down the sharp edges of the glass and collect on top, forming a thin layer of dirt. This can turn to mud during a rain, dry, and harden, partially defeating the purpose of the glass. When glass is used, the area must be inspected every three to four months for replacement.

Barbed wire can be intermixed along with the glass on top or a concertina type of barbed wire can be permanently affixed to the wall top.

Buildings

There will be those occasions, and this is more true in private industry than government, when building walls form a part or all of the basic perimeter of the facility. Buildings, or similar structures, that are used instead of a standard type of perimeter barrier or as *part of* the fence line perimeter barrier, must provide a protection equivalent to

Table 3.2 There are a number of advantages of using brick walls over using chain-link fences for perimeter security

Advantages	Disadvantages
More stability and strength	Requires a firm foundation
Cannot see through	Much more costly than chain-link fence barriers
Height can be increased (to some extent)	Limited height factor
Variety of toppings can be included	

that of the fence (or wall) required for that area. This means that all windows, doors, ventilation openings, ceiling access holes, or other means of access must be guarded or properly secured against the potential intruder.

Whenever a building wall forms a part of the perimeter, a protective grillwork, laminated shatterproof glass, or other type, such as solid block glass, should be used to cover windows. If possible, the windows should be permanently closed, preferably by bricking.

Doors and/or other openings along the perimeter barrier should have a protective grillwork installed or be permanently closed and bricked.

The floors, roofs, and so forth of such buildings, and various ducts, vents, and/or other similar openings will also have to be protected with expanded metal coverings, bars, or alarm systems, with each protection point being determined from an individual evaluation. Some of the specifics for these types of openings and their proper protection are covered later in this section.

Generally, openings can have a minimal amount of protection (depending upon facility sensitivity) if they are more than eighteen to twenty feet above the ground or at least that distance from a roof point that is located directly below or immediately adjacent to and below the opening. Also, any opening with an area of ninety-six square inches or more that is less than eighteen to twenty feet above ground *or* less than fifteen feet from an uncontrolled structure that is *outside* the perimeter barrier must be secured with bricks or bars.

Windows

During the past ten to fifteen years, the architectural trend has been to create wide open spaces with lots of windows. To a great extent, this flies directly in the face of good physical security practices. The use of glass and light may be good for employees, but can become a continuing headache for security personnel.

In general, there are several methods by which adequate protection can be achieved for such windows, including:

- Grills
- Bars
- Special fastenings

- Special glass or laminates
- Installation of window sensors, room sensors, and CCTV
- Use of draperies and covers

Grillwork is common within private industry, and in buildings and areas that are contracted out, rented, or purchased by the U.S. government. Grills may take on an ornamental design, and the design development can provide security weaknesses or positive points, depending upon the amount of attachment points and methods of tying it to the building, its flexibility, the thickness of the solid grill pieces, and so on.

Grills can consist of simple or complex designs that indicate an organizational logo or initials, or they can be very basic and look like the bars that they are. Long, heavy screws or bolts attach the grills to the building or wall. Some grills that are installed during the building process are affixed into the structure, with the grill cross-members going *into* the actual structure, usually between bricks. Grills have been used, at times, to cover or hide blemishes on a building's surface, but mostly are used to cover small and medium-sized areas, such as doors or windows. Sometimes a movable grillwork is put up over an entranceway but is used as a part of the security barrier only during hours when the building is closed.

Whenever possible, grills should be set back from glass at least four to six inches. They should be affixed with carriage bolts, the bolts being at least four inches in length. The metal portion of the grill, when not welded on both sides, should also have a round-headed bolt running through it. Although movable grills are available, consider using permanent grills only. Flexible-type movable grills, especially those that are drawn across large entranceway areas, are too flexible in terms of sway.

This sway factor can be used to an intruder's advantage. An intruder may be able to pry up and out, or down and out from the top to gain access.

If a flexible grill has been installed, make sure it has top, bottom, and side running grooves, with a series of steel vertical rods and interlocking cross-bars within the unit. The grooves at the top and bottom allow the grill to "lock" into them and prevent the grill from being moved outward and bent at an angle. The side grooves allow the grill to "ride" within a controlled track, if necessary; they also serve the same function as the top and bottom grooves. The steel rods and interlocking cross-bars also help reduce the sway factor to a minimal level.

Always keep in mind that this type of movable grill provides only an average amount of protection, and its deterrent value to the determined intruder is very, very low.

Further, note that the cost of flexible grills is somewhat high, and these grills are usually made only in standard sizes. A special order will be expensive, and the security benefit will not be as great as another method.

Protective grills—such as those covering glass doors and window areas—can also be made from welded mesh or expanded metal. They are usually frame-mounted to a continuous edging of steel and bolted into place. The bolts should be round-headed on the exterior side and pass entirely through the frame on the interior side. The nut/bolt on the interior side and the round-head on the exterior side should be spot-welded.

Protective metal grills with a continuous metal hinge (called a "piano hinge") on one side allow for emergency exiting by personnel. Such grills are locked on the interior side; the key is maintained within the room, at least 40 inches from the inside edge of the grill closest to where the key will be located. This distance ensures that someone cannot cut through the metal, break the window, and then reach in to get the key and open the lock.

In this case, the grill should be bolted to the wall on all sides of the grill frame. The gauge of metal should be commensurate with the risk level determined for the facility, and it should be able to withstand a given level of attack.

Note that expanded metal is *far* superior to welded or a standard wire mesh grill. Expanded metal is made by using a solid, singular piece of metal that is then cut and expanded in a cold state. Because it is not welded, jointed, or pieced together, it is exceptionally strong and resistant to normal attack methods.

Ornamental grills are found more around commercial structures and have a limited security protection. They can be made of wrought iron or other metals, and they are usually only 1/8 to 3/16 inch in thickness, while their width is about 1/2 inch. Because they are chiefly used for decoration, the security protection level must be considered relatively low to moderate, depending on the method used to affix them to the overall structure. Unfortunately, they are usually attached with small bolts or screws up to 1 1/2" long. The grills do not have cross-members and are not usually affixed *within* structural walls, but into the wall by the screws. Bolts are rarely used to hold such grills unless specifically requested when the grills are installed.

Bars should be used whenever possible. They provide a much greater degree of security protection than standard, commercially available grillwork. Bars should be set back at least two inches from the external surface of the wall. They should be grouted into the masonry at the top, bottom, and sides to a minimum depth of at least three inches. For bars longer than two feet, use cross-bars or ties every fifteen to twenty inches that are welded at the intersections with the bars.

Bars should be of a mild or better grade of steel, a minimum of ¾ inch diameter and fitted on five- to six-inch centers if possible, with a maximum center of eight inches when cross-bars are used.

Do not use iron pipe for bars because it is easily cut. Bolt cutters can cut through iron pipe rapidly, efficiently, and quietly.

Never use hollow tubing painted with a flat black paint to simulate steel bars! Some commercial establishments that have tried this method have regretted their decision.

Where window frames and/or buildings are constructed of wood on the exterior walls, bars should be cross-welded and bolted into the framework with long carriage bolts. The bolt head on the exterior should be filed down on the edges to prevent attempts at removal. Each bolt should be double-nutted on the interior side. If possible, the bolt head should be countersunk and covered, and the double nuts should be tightened against each other and spot-welded.

Such bolts should be located every eight to twelve inches depending upon the overall area and weight of the bar unit. Bars that are welded to a cross-member, which is then secured to the woodwork, provide a greater level of security than bars without cross-members.

There is such a large diversity of metal window frames—mostly made from extruded aluminum—that no standards or specifications can be enumerated to cover all possible contingencies. For bars that will be secured to such windows, remember that the frames are aluminum, which can be cut or drilled quickly and effectively to remove the bars. When this is the case, the only alternative is to attach the bars to the building with bolts and cross-bars.

Roof windows and/or skylights are found in a variety of names and types. Their size depends upon the purpose for which they were originally installed. In more modern cases, they have been put in well after the building construction. These are modular-type units that are installed after a hole is cut in the roof. They are usually attached on the inside with heavy-duty screws, long nails, or bolts.

Custom-made windows that have security applications are available for buildings. When examining such windows you can ensure that they are bolted down on the inside, the frame is not of cheap metal or wood, or that they have a metal protective casing that is firmly affixed to the building room. Look at the opening to see if it will allow a person to break through to gain access or, if it has an expanded metal grill or bars firmly attached, check to see that they are on the *inside* of the window area.

Because it is difficult to obtain just the right mixture of security and openness when considering skylight-type windows, have them permanently closed off whenever possible.

You may have to use point protection; namely, a specific sensor attached to the window at a point where it opens and a volumetric sensor within the room to which the window leads.

Some modern windows are made of an acrylic plastic, usually about ¼ inch thick. During periods of cold weather, they can be easily cracked and broken. If heavy-duty tape is used by an intruder, he or she can cut down on the amount of noise and debris when the window is broken. Also, a small torch will heat and melt the plastic without the noise usually associated with breakage.

Shutters, where used, should not be entirely of wood. A steel shutter hinged on the interior side and bolted to the frame/wall is best. For aesthetic reasons, the exterior can be covered with a wood laminate to allow the shutters to blend with the rest of the exterior, giving a more natural look.

Whenever possible, consider using a metal cross-bar with a locking device on the inside of the shutters, instead of just a lock and hasp unit. This prevents the shutters from being forced open with a pry bar at the point in the center where the hasp is located. Because the hasp is the weakest link in a shutter system, use a metal cross-bar instead.

Glass areas, or windows of any type, pose a security hazard that must be balanced against the security risk that the facility is willing to incur. When buildings are rented or purchased, consider the advantages of blocking up windows whenever possible. By obtaining the owner's permission, or having it written into the contract when renting, the more critical windows can be blocked up and you can use heavy-duty security screens over other windows. If the windows are of the picture-window variety and the building is rented, you probably will not be able to do this.

When possible, consider the use of wire mesh glass. This will prevent the glass from shattering when it is struck with an object. This type of safety glass may crack or splinter, but it will not shatter under average attack methods. The use of a vehicle or small bomb will, however, eliminate any advantage of wire mesh glass. Wire mesh glass also provides a certain level of protection from fires by reducing the hazard from flying glass. Wire mesh glass is not found as much in government or private industry as it was ten or even twenty years ago, but it is still found in great measure in the true industrial realm.

Laminated glass provides an increasing level of security and is becoming more and more common in today's government and business worlds. Laminated glass has a very thin layer of transparent plastic sandwiched or sealed between two layers of glass. This construction provides a much higher resistance to attack and breakage than standard glass.

Bullet-resistant glass and plastic carbonates are used mostly in high-threat and crime areas, banks, jewelry and related businesses, plus many government offices. It is extremely expensive, and its weight can be four or even eight times more than ordinary glass. It is constructed of several layers of bonded glass interspersed with plastic. Its thickness and the combined attributes of the plastic within the glass determine the resistance to bullet penetration.

Glass blocks, either six-inch squares or nine-inch squares, can be used in lower levels of buildings for office and basement windows. They offer a reasonable level of physical security, while allowing moderate amounts of light to pass through. An intruder will take a lot of time and make a lot of noise attempting to forcibly break the glass blocks.

Roof Access to Buildings

The problem of roof access is not covered in many security inspections of facilities for the simple reason that inspectors don't necessarily think to check out the roof areas. If it can't be seen from the outside ground level or from a rise a short distance away, it becomes like the adage, "out of sight, out of mind."

Consider the fact that all buildings have a roof. There must be some way to get to the roof, either from the inside or from the outside. For many commercial establishments, there will be both interior and exterior methods to gain access to the roof. Because roofs are not normally observed by security guard forces, and because an alarm device or CCTV cameras are not maintained for observing the roof, it becomes an ideal area for an intruder. The intruder can work with relative freedom, worrying only about noise that might be heard by the security force or someone else who would alert local security forces or police. Consider the following points during the inspection phase:

- Is the building located adjacent (within ten to fifteen feet) to other structures?
- Is roof access possible for an intruder from this adjacent building, if the intruder jumps or uses a board, ladder, or other device?
- Is the adjacent building higher or lower than the roof; what is the height difference?
- If a low-storied building is adjacent to a high-rise building, what windows on the adjacent building could be used for access points? Could the building have wire fencing with sensors attached or CCTV to prevent such unnoticed access?
- Are the building walls so high that they will hide intruders on the roof, so that they will not be observed from an adjacent building or from a rise a short distance away? Could they be seen from CCTV cameras mounted in unique viewing positions?
- Are there fire exit escape ladders on the building side walls? Do they extend all the way to the roof and to the ground? Is there a cutoff point for climbing up that would limit a person's access to the roof from the ladder, that is, is the ladder a "one-way" ladder?
- If ladders are available on the premises, are they chained or locked up so an intruder cannot use them to reach a fire escape and then the roof?

These questions will have to be addressed during the inspection by a visual reconnaissance of the building, looking at the building(s) within the facility from the ground, from other surrounding buildings, from a rise a short distance away, and from the areas of the roof itself.

Other Access Points

Older buildings, or those buildings that are built strictly for industrial use, or those within downtown business areas have several items that are not normally considered when evaluating the overall facility physical security posture.

Older buildings may have old, unused coal chutes, or even tunnels between buildings, bricked-up areas between the basement of the inspected building and a building next to it. In addition, other buildings can have manhole openings beside the building (sidewalk area) or directly adjacent in the street, providing access to tunnels that run very close to, or under the building. Grates, ducts, and elevator shafts, as well as telephone and local electrical company and water line utility passageways that go through, under, or near facility buildings should be examined. Security concerns must always include new and alternative methods to see if the security can be breached. After all, that is what the intruders are looking for.

Because the perimeter barrier is the first level of facility protection that must be breached, the intruder will be looking at it very carefully; it is wise that you do the same.

4

Protective Lighting and Utility Services

PROTECTIVE LIGHTING

Protective lighting is the means of maintaining a predetermined level of visibility protection during hours of darkness that is equivalent (or nearly so) to that maintained during daylight hours. It is of considerable value as a deterrent to intrusion and is an essential element of the facility's integrated physical security protection program.

General Requirements

No two facilities are exactly the same, thus a careful study should be accomplished to determine the best visibility practical for security functions such as the identification of personnel, identification badge checks, vehicle checks, the prevention and/or detection of intruders outside and inside facility buildings, and the inspection of unusual or suspicious circumstances that may occur during hours of reduced daylight or during nighttime.

Security lighting, while a security deterrent in a psychological sense, is not used just for this purpose. On a perimeter fence line, for example, it is used where a fence would be under either periodic or continuous observation. It *may* be unnecessary where the perimeter fence is protected by a central alarm system and other lighting is available.

Security lighting is very desirable for sensitive areas or structures within the perimeter line such as vital buildings, storage areas and warehouses, vulnerable control points of communications systems, power supply sources and transfer points, supplies and tool storage, and other areas.

Within interior building areas, an adequate level of lighting allows for the detection of unauthorized personnel who approach general, controlled, restricted, or other sensitive areas.

Emergency lighting system applications must also be considered for general facility hallways, near doors where sensitive materials and/or functions are maintained, near critical points within the facility confines that concern general utility support for the facility, and also for security posts and centrally manned stations.

Characteristics of Lighting

Lighting allows for a reduced security force within the facility and also provides some semblance of personal protection for security guard force personnel by reducing the advantages of concealment and surprise. Lighting can be less intense than in working areas, except for the identification and inspection at various authorized entrances and also in emergency situations.

Each area of the facility must have its lighting level based on the physical layout, terrain, and protective requirements that are necessary to maintain a predetermined level of security for that area. The lighting must operate continuously during periods of reduced visibility, and standby lighting must be maintained and periodically tested.

Principles of Protective Lighting

For security protection, protective lighting should:

- Enable security guard force personnel to observe activities in a given area without disclosing their presence to individual(s) who are being observed.

- Be used *in conjunction with* other protective measures such as fixed guard posts, foot patrols, fences and other barriers, and alarm systems.
- Be achieved by adequate levels of light on bordering areas; the light is directed outward into the eyes of a potential intruder. Limit the amount of light on security patrol routes.
- Allow security force personnel to see below the level of contrast.
- Ensure a high contrast between the intruder and the background; this is the first consideration of light level determination. When the same amount of light falls on an object and the background, an observer depends upon the contrast between them. For poor contrast, increase the illumination level.
- Eliminate shadows. Dark areas are the greatest ally for a potential intruder.

Employment of Lighting

Various types of lighting equipment are available and the methods of employment can provide an optimum level. The specifics for the employment of lighting in various situations are as follows:

1. *Perimeter (Glare Projection Method)*. When it is not necessary to limit the width of a lighted strip outside the perimeter boundary fence (due to the annoyance of the light), it will provide a strong deterrent, making it difficult to see inside the property line. In this situation, the use of high-pressure sodium-type floodlights is preferred by many.
2. *Perimeter (near adjoining property)*. Where it is necessary to limit the width of the lighted strip outside the perimeter fence in consideration of adjoining property owners, controlled illumination is required. The normal solution is to apply standard street lighting instruments. It should provide a symmetrical distribution of light along the fence line when the lights are mounted vertically and tilted to light a narrow strip inside the fence and a wider strip outside.
3. *Perimeter (near adjacent buildings)*. In some areas where buildings are near the street, the existing street lighting may be sufficient. Where intruder entry can be effected from a nearby

building, consider installing a light-weight fence and/or a light on the face of the facility building.
4. *Entranceways*. The lighting problem at entrances includes the inspection of identification badges, the visibility into truck cabs and automobiles, the inspection of trucks and freight, and the illumination of areas surrounding the gate house. The proper inspection of identification badges requires a higher level of illumination than the average street light illumination provides. Where temperatures do not fall below zero, fluorescent units are preferred because they are brighter than standard incandescent fixtures. Seeing into truck cabs and automobile interiors requires that light be directed so that a visual inspection can be made of the individuals and interior areas, but the light should not leave the interior in shadow. Care must also be taken to avoid excessive glare—a hazard to other local traffic. The interior of the guard post should be kept slightly darker to give the officer on duty a better view outside. Keeping the guard post darker also affords a small measure of personal protection.
5. *Locked Gates*. Attempts at entry are often made at locked gates, even in preference to a more remote location on the facility perimeter. Extra light should be provided by spacing fence lighting units closer together or by simply adding two or three small floodlights to overlap the locked gate. When the gate is used occasionally at night, lighting should be comparable to that used for a fully active entranceway.
6. *Parking Lots*. Parking lots located inside facilities should have a lighting level equivalent to that of standard street lighting. There should be no area of the parking lot or lines of access (including walkways) that do not have adequate lighting. In addition to general security, safety of personnel walking to and from their vehicles must be considered.
7. *Outbuildings*. Such buildings can include supply, general maintenance, warehouses, and utilities. Street level lighting should be maintained around entranceways and doors, between buildings where an intruder could hide, near open storage sites, and other areas where darkness and shadows limit viewing. Access gates into controlled areas should also have this level of lighting maintained during periods of darkness.

8. *Building interiors.* Within building structures, lighting should be maintained at least at 75 percent of that used during daylight hours for hallways. For door entrances going into the facility, the lighting inside and outside should be the same as during daylight hours. For entrances that allow access to restricted or controlled areas, those used for open storage of materials, and areas containing sensitive or government classified information the lighting level should be similar to that used during daylight hours. At no time should hallways be left in darkness. A minimum 50 percent light visibility must be maintained. This is accomplished by keeping half the hall lights on.

Keep in mind that thieves and other intruders do not like light. Because lighting partially restores the protection normally available during daylight hours, it is essential that security personnel ensure that all areas of interest and concern are adequately lighted.

The nighttime levels of light in alleyways between buildings, for entrances, fire escape ladders on sides of buildings, and ground-level windows, must be sufficient to enable security force personnel to detect an intruder who is attempting an unauthorized entry, as well as the mere presence of the intruder.

Lighting is a deterrent only when it is used properly and with sufficiency to provide a psychological control factor.

Emergency/Backup Power

All facilities should have a backup power capability. This can be a secondary external power source, a generator or series of generators depending upon facility area and amount of emergency lighting required, or battery-powered emergency lighting systems.

Any such system must be checked randomly, but should be checked at least monthly. Testing on a regular schedule is important for detecting faulty equipment, burned out equipment or light bulbs, and for ensuring that equipment has not been deliberately tampered with.

The system(s) in use must ensure that sensitive or restricted areas within the facility, main entrance access points, the guard posts, and the intrusion detection system continue to maintain the required security level.

Emergency power sources, whether a generator or battery system, or even a wholly separate secondary power source for the facility, should provide a minimum of six hours of emergency light.

In the case of generators, this means the facility must maintain an adequate supply of fuel. Batteries for backup power sources should be checked to ensure that their fluid level is maintained.

When other power is supplied from an outside source, the power lines should be monitored visually to ensure there is no disruption from human or natural forces.

Other Utility Services and Concerns

The importance of utilities and services during normal working hours, and certainly during an emergency, cannot be overstated. The disruption of any or all of these utilities can seriously impair the ability of the security forces—not to mention the management—to protect, control, and/or provide certain safety procedures during an emergency. Thus, they must be properly protected. Coordination should be effected with the utility service companies to ensure continued service at all times.

Electric power is of paramount importance to the facility. You should determine where the power lines enter the facility, whether there are additional transformers within the facility, and where the lines enter from outside the facility. Are there alternate distribution lines available, within a reasonable distance outside the facility, that can be used in extreme emergency situations? Can the power company tie in to them rapidly or can a secondary line be run to the facility and maintained within a switching mode to allow for changeover if necessary?

If these areas can be covered, in addition to ensuring an acceptable level of physical security protection for transformers and lines within the facility, you have covered the main concerns.

Regarding emergency/backup power, refer to the previous section. In general, know the size and location of any backup generators, their current fuel supply, and who are the designated operators for such equipment. Flashlights and general purpose lanterns should be available in the security office, at guard stations, and interior rooms. Batteries should be checked at least quarterly and replaced when necessary.

There are also concerns in the area of fuel sources, such as pipelines that run to, or through, a facility. These include diesel fuel, oil, storage tanks, and other combustibles. The adequacy of any stockpiles should be monitored. Evaluate where standpipes and control wheels are located, where pipes run underground and the depth, where pipes are open (inside and outside of facility buildings), and where pipes or storage facilities are close to roads (within and outside facility grounds) in terms of physical security protection and safety for occupants and nearby buildings and/or residences.

5
Types of Security Areas

PRINCIPLES OF SAFEGUARDING

All safeguards for the protection of information, materials, and design must be developed so that this material, information, and equipment is protected from inadvertent disclosure to any unauthorized individual and from any deliberate attempts to gain access by either forced or surreptitious entry. Areas of concern include unclassified, sensitive unclassified, and government classified information; controlled access areas and, for equipment, design, production and process techniques; company private materials and technologies, including strategic planning information.

Such safeguards *must* be planned and developed within the concept of security in depth. Depending upon the amount and sensitivity of material involved, safeguards should incorporate an appropriate combination of physical security protection characteristics, namely, access controls, alarms, construction, segregation or compartmentalization of certain activities or elements, and secure storage containers or areas.

Critical and Restricted Areas

In considering the inspection process to this point, only the broadest aspects of building security have been considered. The amount of time and money required to secure the facility may not seem to be wholly justified, thus PSI team members must now start to identify those areas within the facility that will require an increased level of security preparedness.

Restricted areas (which may also be referred to as security controlled or limited access areas) within a facility require a greater level of protection than the perimeter barriers. These areas must be identified and provided a higher security level, as warranted by the functions and/or activities occurring within such areas.

Confer with management to see if they have determined what areas are the most critical and vulnerable. Next, determine if the evaluations of these areas are up-to-date with regard to any changes in facility missions and functions, responsibilities, and levels of interaction with other facilities and organizations. Management should be aware of the general vulnerabilities of the facility in terms of geographical location, facility construction, equipment on-site, any degree of material combustibility, and the ability to extinguish fires should they occur.

The number, size, and location of critical, controlled, restricted areas should be assessed. Such limited access areas can include cash areas, air conditioners, heating, major electrical points (including transformers and electrical switching), main telephone closets, and the like.

That degree of restriction is the next area where questions may be asked. Should access be limited and, if so, what degree of access limitation will be required? How will access be controlled to these areas, and how is access controlled to areas that allow further access to the critical or restricted area?

If an access list is to be used, who will determine what individuals will be authorized access, how additions or deletions from the list are made, and what backup documentation will be required? In the absence of a primary individual, who will be the alternate for adding or deleting names?

If alarm devices are used, are there other backups, such as roving or fixed patrols or posts, CCTV, and other devices?

Once these areas are determined and the methods of direct access and surveillance over access points are set up, there should be a complete check of the actual physical barriers that surround the area. The inspection will include such items as walls, ceilings, floors, ventilation system points, locking devices, key control, and package control.

Although key and package control may not seem to be related to physical barriers, keep in mind that a package can contain a bomb that may damage or destroy a barrier—to say nothing of the materials protected by the barrier—and a key can be used to pass through a barrier. These issues are discussed in greater detail elsewhere in this handbook.

Security-Controlled Areas

All areas within an activity facility must be assigned a security area designation. Different areas and tasks involve different degrees of security depending on their purpose, the nature of the work performed, and the information and/or materials concerned. Similarly, different areas within an activity may have varying degrees of security importance. A careful application of restrictions, controls, and protective measures appropriate to these varying degrees of security importance is essential to good physical security. In some cases, the entire area of an activity may have a uniform degree of security importance, thus requiring only one level of restriction and control. In others, differences in the degree of security importance will require further segregation for security interests.

Areas must be designated as either restricted or nonrestricted. Restricted areas are those that have controlled access. Specific limitations and authorities must be established in writing. For U.S. government facilities, or for contractor facilities under a government contract, the establishment in writing is "pursuant to lawful authority" as promulgated pursuant to Section 21, Internal Security Act of 1950.

The Department of Defense (DOD) and/or government contractors should review DOD Directive 5200.8 (or other government agency equivalent). DOD provides the best standards; other government agencies have tended to follow DOD's lead in this area.

In some cases, the facility or agency head may have the authority (under applicable laws and statutes) to declare an area restricted, thus limiting access and defining the local controls that will be applied to that area or facility portion.

Within government, there are three levels of security sensitivity for different areas. In a descending order of importance, they are: exclusion area, limited area, and controlled area. These designations meet differing levels of security sensitivity while still providing an effective degree of protection, varying degrees of access, and movement control. The following provides a more definitive explanation of the areas.

1. *Exclusion area.* An exclusion area is one containing classified information; access to the area constitutes, for all practical purposes, access to the classified information. The following security measures will be required, at a minimum, for all exclusion areas:
   - A clearly defined perimeter barrier.
   - A personal identification and control system.
   - All points of ingress and egress either guarded or secured and alarmed.
   - Admittance only of persons whose duties actually require access and who have been granted an appropriate security clearance.
2. *Limited Area.* A limited area is one containing classified information; uncontrolled movement in the area would permit access to classified information, but such access may be prevented by either escort or other internal controls and restrictions. The following measures are required, at a minimum, for all limited areas:
   - A clearly defined perimeter barrier.
   - A personal identification and control system.
   - All points of ingress and egress guarded, controlled, or alarmed.
   - Appropriate security clearances held by all persons with freedom of movement within the area. Persons not cleared for access to the information contained within the limited area may, with appropriate approval, be escorted into the area and must remain under *constant* escort until they leave the area.
3. *Controlled Area.* A controlled area is one in which uncontrolled movement does not permit access to classified information; it is designed for the principal purpose of providing administrative control, safety, and as a buffer zone for a limited or an exclusion area. The following security measures are required, at a minimum, for a controlled area:
   - A clearly defined perimeter.
   - A personal identification and control system.

- All points of regularly used access maintain a station where personnel identification *may* be checked.
- Security and administrative arrangements have been established to determine the need for entry and method of approval for admittance to the area.

All areas that do not fall under the three restricted areas described above are referred to as *nonrestricted areas*. A nonrestricted area is one that is under the control of the facility, but access to it is either minimally controlled or uncontrolled. Such an area may or may not have a perimeter definition and may be open to the uncontrolled movement of the general public at times. An example of a nonrestricted area is a visitor or employee parking lot that is open and unattended by guards during business hours. After normal business hours, it may be closed, patrolled, and converted to some form of controlled-access area. A nonrestricted area could be an area enclosed by a fence or other barrier, to which access would be minimally controlled by a checkpoint, which would only ensure that the person gaining access has official business or another authorized purpose. In such cases, further authorization would not be required. However, once the individual approached a building, other access controls may be in place. Nonrestricted areas are never located within a restricted area.

6

Construction Standards and Requirements

Within facilities, construction of restricted areas that require a limited access is a function of the security office. As part of the PSI, team members should be aware of standards and requirements of such construction. Security concerns start at the time it is determined that construction of a restricted area is necessary within the facility. The planning, development of contract requirements, observation inspection and review of ongoing construction, and in some cases, the testing of the newly constructed area upon completion, are overseen by the security staff of the facility. Team members should check the various construction phases when observing an area under construction, know whether or not minimum standards and requirements are being met when viewing a security upgrade, and evaluate a completed area to see if all the basic requirements have been met.

In addition to these specifics, perimeters of individual buildings within a facility have certain basic security standards and requirements that should be met to assure an adequate protection level for the individual building and the entire facility. The areas of involvement for basic concerns are detailed on the following pages.

BUILDING ACCESS POINTS

Exterior Doors

Each door should be as formidable as possible, considering the aesthetic value and location of the door. The very front door to the facility will not normally be of solid wood. For all outward opening doors that have removable exterior pin hinges, the pin hinges must be replaced with nonremovable pin hinges or else the removable pin should be made fast by welding or inserting a set screw that will "lock" the hinge to the pin. The screw should not be accessible from outside the door. The exterior hinge leaves should be firmly affixed to the door with bolts such that they cannot be disassembled from the outside by a potential intruder. Rear and side door openings that lead to an alley or other secluded area, as well as roof doors to stairwells and elevator housings, should be resistant to lockpicking and other basic forms of attack. Where possible, windowless metal-clad doors are recommended, or else solid wood. Door ventilation openings must be secured in place with a series of metal bars over the opening to prevent access. Whenever possible, these openings should be replaced with solid doors.

Exterior Accessible Windows

If possible, all side and rear windows near secluded outside areas should be barred or covered with a heavy-gauge steel screen, permanently bolted into the wall. If windows are not really required, brick them up with solid, regular bricks or glass bricks. Windows should be kept locked at all times, except when an employee is in the room.

Roof Openings

Roof openings include transoms, skylights, ventilator shafts, air ducts, and the like. If an opening is greater than 96 square inches, it should be covered with heavy-gauge steel screening or barred. Roof covers should be metal or metal clad, and there should be no exterior accessible removable hinges, pins, or locks. Expanded metal can be put on the interior side of currently installed covers. Other openings, such as louvers, should be secured with bars or heavy-gauge steel screening. A point sensor alarm located on the interior side of the opening can also be used.

Fire Escapes

The external fire escape to a building often gives an intruder a method to enter the upper building stories or roof area where he or she can work unhindered. Doors leading to the fire escape should be one-way, opening outward, with no locking device assembly keyway or handle on the exterior side. There should only be a locking or other release device on the inside to maintain the integrity of the door as a barrier. In these cases, the "crash bar" or "panic bar" type of exiting feature should be employed to comply with appropriate fire codes and regulations. Whenever possible, buildings should employ an internal fire escape (stairwells). However, such stairwells can also be used to enter and move upward through the building. Fire regulations require that stairwells be open—but because they are for exiting rather than entering the facility, the access should be regulated by security policy and procedures.

VAULT CONSTRUCTION STANDARDS AND REQUIREMENTS

These standards furnish recommended requirements for the design and construction of vaults intended for the storage of classified records, information and material, or other sensitive or high-value items that require protection (see Table 6.1).

Elements of Construction

Except as otherwise noted, the design for the walls, floors, and roofs (ceilings) of a vault should be in accordance with nationally recognized structural engineering standards. The thickness should not be less than specified. Where vault walls are contiguous to the exterior wall of a facility building, the vault wall(s) should be set back from the exterior face of the outside wall(s) to allow at least four inches of the normal wall facing material to cover the vault wall(s).

Concrete used in vault construction should be monolithic, poured and cast in place, with a minimum compressive strength of at least 2,500 psi after twenty-eight days.

In terms of safety and emergency devices, the doors to vaults may or may not be equipped with an emergency escape device. However,

Table 6.1 Minimum construction standards for class A, B, and C vaults

Vault Classification	Approved Storage Level	Thickness		
		Floor	Walls	Ceiling
A	Top Secret	8" RC[1]	8" RC	8" RC
B	Secret	4" RC	8"[2]	4" RC
C	Confidential	4" C[1]	8"[3]	4" RC

RC = Reinforced Concrete

C = Concrete without reinforcement

NOTES:

1. All concrete used in vault construction will be monolithic cast in place, conforming to a minimum compressive strength of 3,000 psi after 28 days of aging. Reinforcing will be by a minimum ⅝ inch diameter steel reinforcing bars (rebar) laid to a maximum of 6 inches on centers, creating a cross-hatched steel curtain, to be sandwiched at half thickness of the concrete, parallel to the longest surface. Rebar will be anchored or embedded in all contiguous walls/surfaces.

2. Class B vault walls will be constructed of masonry at least 8 inches thick, such as brick or concrete block employing adequate bond. Hollow masonry, only of the vertical cell (load-bearing) type, can also be used, but if used, each cell will have from ceiling to floor ½ inch diameter or larger rebar inserted and then be filled with pea gravel and Portland cement grout. Rebar will be anchored in both floor and ceiling to a depth of at least 4 inches. In seismic areas, 6 inches or thicker RC will be required.

3. Class C vaults will be constructed of thick-shell concrete block or vertical cell clay tile and be not less than 8 inches thick. In areas of somewhat frequent seismic activity, 6 inches or thicker RC should be used.

should a vault door not have an escape device, the vault interior must be equipped with an interior alarm switch or device—such as a telephone, radio, or intercom—to permit a person locked inside to obtain assistance.

Vault doors do not have a specific fire rating, so the vault interior should have an automatic, wet-pipe sprinkler system (conforming to the National Fire Protection Association (NFPA) Standard No. 13) with a water-flow device to sound a local alarm and transmit a signal when the sprinkler system has been activated.

Heating and ventilation duct openings must be kept to the minimum actually required. Where heating and ventilating supply and the air return must pass through the vault walls, floor, or ceiling, the air stream should preferably be carried through one or more circular openings (less than eight inches in diameter) that have been cast into the wall. If such an arrangement is unworkable and rectangular openings must be used, then the openings should not exceed six inches at the narrowest dimension. Connecting ducts should be rigidly attached to the vault exterior by riveted connections to an angle iron frame held in place by expansion bolts set in the concrete. The bolt heads must be tack-welded or otherwise treated to impede removal. Preferably, all duct connections should be in exposed locations where they may be readily observed.

Heating and ventilation ducts should be kept to an absolute minimum in areas where classified material, sensitive unclassified, or other valued items are stored on open shelves. Where the supply and return ducts must pass through the vault, the air stream must be carried in a pipe cluster barrier. This barrier must consist of steel pipes not larger than three inches in diameter or more than four feet in length, and in sufficient number to carry only the requisite air flow. The pipes should be in clusters, square or rectangular in cross-section, and rigidly welded together. No. 3 reinforcing bars can be welded to the four sides of the cluster at approximately six inches on center to extend into the vault wall for at least six inches. The cluster is centered in the wall longitudinally. Connecting ducts outside the vault wall(s) should cover the projecting ends of such clusters and be rigidly attached to the vault wall by riveted connections to an angle iron frame held in place by expansion bolts in the concrete. Bolt heads are tack-welded or otherwise treated to impede removal. Duct connections should be in exposed locations where they may readily be observed.

Pipes and conduits entering the vault should pass through the walls that are not common to the vault and the structure housing the vault. Preferably, such pipes and conduits should be placed before the vault is cast. If this is not practical, then they must pass through snugly fitted pipe sleeves cast into the concrete. After installation, the annular space between the sleeve and pipe or conduit must be caulked solid with lead wool. All pipes and conduits inside the vault should be exposed on the interior surfaces.

Class "A" Vaults

Floors and Walls Floors and walls are constructed of reinforced poured concrete not less than eight inches thick. The walls must extend up to the underside of the roof or floor slab above and must solidly connect. They shall be reinforced with reinforcing rods, at least ⅜ inch in diameter, mounted vertically and horizontally, on center between two and ten inches.

Roof The roof should be of a monolithic, reinforced concrete slab of a thickness to be determined at the site, but not less than eight inches thick.

Ceiling The ceiling—where the underside of the roof slab or roof construction exceeds twelve feet in height, or where the roof construction is not in accordance with the roof specifications previously given—requires a normally reinforced concrete slab to be placed over the vault area at a height not to exceed nine feet. The structural ceiling will be at least equal in strength to that of the vault walls.

Vault Door and Frame Unit The vault door and frame unit shall afford the protection at least equal to that of a Class V vault door.

Lock and Locking Parts The vault door combination lock shall conform to the U.L. standard for a Group 1-R three-position combination lock and have a top-reading, spy-proof dial. The U.L. label is considered adequate proof of compliance with this requirement. When properly installed, the axial play on the lock handle spindle shall not exceed 1/16 inch. The lock, locking bolt, door bolt operating cam, and bolt operating linkage connected thereto shall be protected by a tempered steel alloy hardplate located in front of the parts to be protected. This hardplate must be at least ¼ inch in thickness. The front plate, edge plates, back plates, and cap sheet shall be of the manufacturer's standard construction. The cap sheet of the door will have an inspection plate of such size that its removal will permit examination and inspection of the combination lock and operating cam area without removal of the entire back cap sheet of the door.

Class "B" Vault

Floor The floor shall be of monolithic poured concrete of the thickness of any adjacent concrete floor construction, but not less than six inches (eight inches would be preferable).

Walls All walls must not be less than eight-inch thick brick, concrete block, or other masonry units. Hollow masonry units shall be the vertical cell type (load bearing) filled with concrete and steel reinforcement bars. Monolithic, steel-reinforced concrete walls at least four inches thick may also be used, and shall be used in any area experiencing seismic disturbances. Whenever possible, and within budget constraints, consider the wall of concrete, and make it at least the same thickness as the floor.

Roof The roof must be a monolithic, reinforced concrete slab of a thickness to be determined by other vault structural requirements, but at least the same thickness as the floor and walls.

Ceiling When the underside of the room slab exceeds twelve feet in height from the floor, or where the roof construction is not in accordance with the above roof specification, a normal, reinforced concrete slab will be placed over the vault at a height not to exceed nine feet. The thickness shall be equal to that of the floor and walls of the vault.

Vault Door and Frame Unit These shall be the same as required for a Class "A" vault.

Lock and Locking Parts These shall be the same as for a Class "A" vault.

Class "C" Vault

Floor The floor is the same as for a Class "B" vault.

Walls The walls are made of not less than eight-inch thick hollow concrete block (thick shell). A monolithic, steel-reinforced concrete wall at least four inches thick may also be used and must be used in areas with seismic disturbances. The walls behind the exterior wall portion of the building shall be concrete, solid masonry, or hollow masonry units filled with concrete and using steel reinforcing bars.

Roof, Ceiling, Vault Door and Frame Unit, and Lock All shall be the same as required for a Class "A" vault.

All vaults must be equipped with an emergency lighting system and an intercommunications device—a telephone, intercom, or radio unit that is capable of contacting the vault supervisor, facility security guard, or local police to assist a person locked inside the vault.

Strong Rooms

Some facilities may only need a strong room rather than a high-level or sensitive, secure storage area. By definition, a strong room is an interior space that is enclosed by or separated from other similar spaces by four walls, a ceiling, and a floor, all of which are constructed of solid building materials. Using these criteria, rooms with false ceilings and walls constructed of fabrics, wire mesh, or other similar materials do not qualify. Such construction can be used for interior areas within a controlled-access facility, used as work spaces where classified or sensitive facility operations are undertaken, used where automated information systems (AIS) terminals are in a hard-wire configuration providing direct contact with a computer mainframe, and used under circumstances in which open storage may be desired for classified or other sensitive types of information. Under such criteria, specific construction standards are as follows.

Walls and Ceilings All walls will be true-floor-to-true-ceiling. Construction shall be of plaster, gypsum board, metal, hardboard, or wood over a solidly constructed frame of two-by-four-inch boards. Plywood or other materials that offer similar resistance to, or provide evidence of, unauthorized tampering or entry into the area may also be used. Insert panels shall not be used. Minimum thickness of the wallboard used shall be ½ inch.

Floors Floors shall be of solid construction, utilizing materials such as concrete, ceramic tile, wood, and similar materials.

Hardware Heavy-duty builder's hardware shall be used throughout the construction, and all screws, nuts, bolts, hasps, clasps, bars, hinges, and pins, shall be securely fastened to preclude entry and assure visual evidence of attempted or actual forced entry. Hardware accessible from the exterior of the area shall be peened, brazed, or spot-welded to preclude removal.

Windows For applications within the scope of this manual, no windows or window areas are authorized.

Miscellaneous Openings Where duct work, registers, tunnels, or other openings are of such a size and shape as to permit unauthorized entry or visual access, they shall be equipped with barriers such as heavy wire mesh (No. 9 gauge, two-inch square) or steel bars at least ½ inch in diameter extending across the width of the opening with a maximum space of six inches between bars. The steel bars shall be securely fastened at both ends to prevent removal and shall have cross-bars to prevent spreading. Where wire mesh or steel bars are used, care must be taken to ensure that material within the room cannot be removed with the aid of any instrument.

Doors Doors shall be substantially constructed of solid wood or wood with a metal layering to the front side.

Door Louvers and Baffle Plates When used, these must be reinforced with a wire mesh and fastened from the inside.

Door Locking Device Locks shall be either a built-in, three-position, dial-type, changeable combination lock, or a three-position, dial-type, changeable combination padlock that is secured to the door by a solid hasp. A combination padlock for these purposes will only be used on the inside of secondary doors to the room. All entrance doors will have the first type lock indicated.

Alternate Vault Construction Standards

An alternative to the Class "A" vault specifications is a vault of steel-lined construction, as specified below. They are for use in an existing or new structure above or below ground level.

Construction is of steel plate, a minimum of ¼ inch thick. The steel plates are continuously welded to supporting steel members of equal or greater thickness. The supporting steel members are placed in a contiguous floor and ceiling of reinforced concrete and must be either firmly anchored to or embedded within the floor and ceiling. If either the vault area floor or ceiling construction is less than eight inches of reinforced concrete, a steel liner will be constructed, the same as the walls, to form a floor or ceiling for the vault. Because a Class V vault door is used only with reinforced concrete walls, the Class VI door will be used on steel-lined vaults.

All vault facilities will have a vestibule constructed at the front entrance with an access door to achieve control when the door is in the open position. The use of a vault door for controlling movement into and out of the vault is not authorized because this continued use will create undue wear on the door and will eventually weaken the locking mechanism, causing malfunctioning.

When building codes require that the vault entrance meet a specified fire rating, the vestibule and its access door must be of the required fire rating. Where permissible, the vault door operational day gate may be employed as the work day entrance control in lieu of a vestibule. Such devices, however, will hamper the use and discussion of classified material and must not be used at facilities in which classified material is either being processed or is available within the vault area.

There will be no windows in the vault, and all ventilator openings or other access routes into the vault will be properly treated to deny unauthorized access.

Security Upgrade Construction Standards and Requirements

The following is offered as a norm against which to evaluate the adequacy of current structural security standards. These standards also provide some guidance relative to new construction within areas of facilities that may contain activities and material that must be protected.

Hardware Heavy-duty builder's hardware will be used in construction. All screws, nuts, bolts, hasps, clamps, bars, two-inch square mesh of No. 11 wire, 18-gauge expanded metal screen, hinges, pins, and the like should be securely fastened to preclude surreptitious removal and to assure visual evidence of tampering. All hardware accessible from outside the area should be peened, pinned, brazed, or tack-welded to preclude its removal. The term "two-inch square mesh of No. 11

wire" meets standards of Federal Specifications and is also referred to as wire mesh.

Walls Construction should be of plaster, gypsum wallboard, metal panels, hardboard, wood, plywood, or other opaque materials that offer similar resistance to and provide evidence of unauthorized attempted or actual penetration into the area. If insert-type panels are used, a method must be devised to prevent their removal without leaving visual evidence of such tampering. (NOTE: For this reason, such panels should not be used unless no other type of material is available within a reasonable time and distance from the facility.) Area barriers up to a height of eight feet should be of opaque or translucent construction where visual access is a factor. If visual access is not a factor, the area barrier walls may be of wire mesh or other nonopaque material that offers similar resistance to and provides evidence of unauthorized entry or attempted entry into the area.

Doors Doors will be substantially constructed of wood or metal. When there are panels within the door, such panels will be removed and the door made solid.

Door Locking Devices Entranceway doors will be equipped with an approved built-in, three-position, dial-type, changeable combination lock.

Ceilings Ceilings will be constructed of plaster, gypsum wallboard material, panels, hardboard, wood, plywood, or other material offering similar resistance to detection of unauthorized entry. Wire mesh, 18-gauge expanded metal screen, or other material offering similar resistance may be used if visual access is not a factor. The ceiling will meet and overlap to the outer edge of the true-floor-to-true-ceiling walls. If there is a false ceiling within the room, the walls must pass it and meet with the true ceiling.

Miscellaneous Openings All such openings will be protected as indicated as for a strong room area.

Alarm Systems All vaults described here—whether of concrete or steel, strong room areas, and facility-constructed areas made for purposes of upgrading—should have an approved alarm system that is monitored at all times. Where a false ceiling is part of the vault or strong room, sensors should also be mounted above and below the false ceiling. The door units must have a magnetic switch mounted on the interior side of the door and frame to indicate unapproved or unauthorized access into the area.

7
Locks, Doors, and Windows

Doors, windows, and their associated locking devices are the primary points of attack an intruder uses to obtain access to a facility or to a specific area within the facility. This section discusses their various elements and features.

You must be aware of the elements for the proper determination of your physical security needs, requirements for planning, their attributes, weaknesses, and places for improvement or replacement. Without this specific knowledge, you cannot adequately evaluate such items to make objective recommendations for any security enhancement requirements for a facility.

Locks are only a part of the deterrent element of a good physical security protection. Every locking device, *given time*, can be defeated. This is either because of an internal weakness—a weakness in the structural environment in which it has been placed—or because of human carelessness. Any of these can be exploited by a knowledgeable or determined intruder. Fortunately, in the majority of cases, time is usually on your side.

Almost every lock can be used as a deterrent. However, it is foolhardy to rely exclusively on the lock. Its application, to be truly effective, must be related to other protective measures currently in force at the facility.

Location, lock rotation, and key control are only part of these protective measures and must be viewed as integral elements of lock utilization. Consider the types of locking devices being used. Ensure that they are "approved" to meet the level of security protection required; the least expensive device is almost certainly not the best. Using the best possible locking devices to suit specific applications is the baseline requirement. This level of protection will be appropriate for that portion of the facility.

During the past ten to fifteen years, the technological advances made in doors and locking devices have concerned security personnel. The door and its lock are the most common points of entry for thieves and/or other intruders. Door locks of all types, especially those with only five pins, *must* be reevaluated with regard to the facility security protection level.

CONSIDERATIONS FOR LOCKING DEVICES

As previously stated, locking devices are used on doors and windows. In determining the appropriate type of lock, its application must be considered. The following are part of the thought process used for this determination.

1. Study all the levels of security for the facility and specific areas where locking devices will be used.
2. The quality of the doors, windows, and frames, to say nothing of the quality of the locks, must be assessed. Lock quality is relatively easily determined, but that of doors and windows is not.
3. Look at the manner in which windows, doors, and associated frames are attached to the rest of the structure.
4. Consider replacement of a door or window, or else reinforcement of the door and window frames and jambs to prevent a potential intruder from ripping, peeling, or spreading them.
5. Consider the fact that case-hardened cylinder guards and steel reinforcing inserts are available for locking devices.
6. The strike-plate for a lock must be secured to the framing with more than a ½-inch or ¾-inch screw that usually comes

with it; better quality high-security locks have 1 ½-inch or 2-inch screws.
7. Observe the tolerance between the door and the jamb; it should never be more than ⅛ inch. A wider tolerance means that the door is easy to open by force without a key; any less, and the door may stick.
8. Check the metal frames; they should be reinforced.
9. Check the hinges to see if they are on the inside or outside of the door. If they are on the outside, consider replacing them or specify that nonremovable hinges be used. Removable hinges can be spot-welded to prevent pin removal.
10. If the door frame is wood, determine whether or not the exterior trim can be removed to allow access to the door locking bolt.
11. Lock cylinders should be solid brass with a full .051-diameter plug. There should be a minimum of six pins to a cylinder, especially if the system uses master keys. In master keying, the system should be kept to the lowest possible level. Each splitting of a master and submaster weakens the overall facility security. As more numerous key combinations become available, the ability to pick a lock at more than one set pattern is increased.
12. Ensure that doors cannot be spread from their frames or pulled open far enough to allow a shim, push-pull knife, or other thin tool to manipulate the bolt.
13. When removable cores are used for a facility locking system, check the type(s) of control(s) for the master keys and the core key. Find out where the cores are stored and whether or not the lock to that area is of the same manufacturer and system as the rest of the system. It might be a lock with an insufficient security level. Find out also if the key has *ever* been loaned to other authorized security personnel who are responsible for the security of the locking system.
14. Key blanks for all facility locking devices should be maintained in the security office. For all facilities, whenever possible, have one type of lock installed, with a specific keyway that is not commonly available on the outside. Manufacturers have designed and controlled certain lock keyways and the associated blanks. They are not normally available to

the average citizen. To obtain and use them, you must go through a factory-approved locksmith or directly to the factory.

One item that must be emphasized is that lock security starts with proper key control. This can never be too strongly stated. The facility key control system should maintain the following information: the key type, key-specific identification, control procedures for signatures and cross-referencing against office/door/area within the facility, the duplication of keys and its control, and the controls on various codes for the keys.

The security storage cabinet for keys, codes, and duplication equipment should be in the security office, and also within any facility-controlled or restricted-access area. Keys, codes, key cutting equipment, and similar items should not be kept in the facility engineer's or administrative offices. Security in these areas is typically at a nominal level.

The highest level of key security should be maintained for those keys that open into the storage area for keys or the associated key-duplicating equipment room. The security guard force should never have keys to these areas. For exceptionally high security areas within facilities, consider using armored face-plates on the locks or steel plate over the entire door surface.

Locks should have an automatic dead latch and cylinder protector rings. Examine the door frames for any necessary reinforcement, especially wood frames or hollow-walled areas.

Hollow metal doors should be of at least 16 gauge, with 14-gauge metal for door frames.

Before discussing the specifics of lock types, it is helpful to have a general background concerning locking devices and systems. The term, locking devices, as used here, refers to key-actuated locks for doors, door-mounted combination locks, drawer-mounted combination locks (for security containers, vaults, and safes), door-mounted auxiliary locks (either push-button or key-activated), key-operated padlocks, and combination padlocks.

A lock system is a device for fastening, joining, or engaging two or more objects. When in the locked or fastened condition, a lock system limits all movement of the objects; in the unlocked or unfastened condition, it permits the movement or separation of the objects. A lock system also includes the means to operate the device so that it is either in the locked or unlocked condition.

Types of Locking Devices

The following provides a short summary of the best available types of locking devices. The specifics should apply to the majority of locking devices selected for a facility. For this reason, this summary does not encompass every type of locking device that could be used but provides a summary of requirements.

Mortise Lock   A fully encased device designed to fit into a square cavity cut into the door edge.

Rim Lock   A case device for mounting on the inside surface of a door.

Tubular Lock   A device installed in a hole bored through the door. The working mechanism is contained within the locking case that is installed through and on the back of the door where it is attached.

Cylindrical Lock   A key-in-the-knob locking device.

Narrow-Stile Lock   A locking device used in narrow-stile doors, usually found in metal-framed doors that surround a large piece of glass.

Concealed Vertical Rod Exit Device   This is a locking bolt concealed inside the door stile with an interconnected deadbolt.

Deadbolt Cylinder Lock   A locking device (either key-in-knob or key cylinder separate) that has a locking deadbolt to prevent easy access by shifting the locking bolt back into the lock. All such devices should have a laminated bolt that is more resistant to sawing. These should be long-throw deadbolts versus short-throw deadbolts.

Interchangeable Core Lock   This lock is part of a larger locking system using locks with a key core that can be removed and replaced immediately by another key core using a different key. Its main features include the following points: (1) cores may be quickly replaced, instantly changing the matching of locks and keys if their security is compromised; (2) all locks can be keyed into an overall complete locking system; and (3) the system reduces maintenance costs and new lock expense.

Conventional Combination Lock This is not a high-security manipulation-resistant combination lock, but one that can be opened by manipulation of the settings of the combination lock wheels. It is a three-position locking device that is not of a high tolerance level in the manufacturing process, indicating that the security protection level is not the highest. Such devices are found mostly in commercial applications where a small security container is used on the premises to store documents, jewelry, cash, and the like; they are also used in residences.

Manipulation-Resistant Combination Lock A manipulation-resistant lock is designed so that the opening lever does not come into contact with the tumblers until the combination has been set. Such a lock furnishes a high degree of protection for important or highly classified/sensitive material. The lock tumblers—the wheels—are made of a polycarbonate material that is less dense than the metal. X-rays cannot be used to determine individual tumbler combination settings. These are identified as Group 1-R combination locks.

Relocking Devices A relocking device on a safe or vault door furnishes an added degree of security. Such a device automatically "spins" the combination lock wheels when the locking device is set into the locked position. This prevents the locking device from being immediately opened by another individual without using the combination. Such a device will appreciably increase the difficulty of opening a lock container by punching, drilling, or blocking the lock or its parts. It is recommended for vaults and high-security containers.

Cypher Lock A cypher lock is one that uses a digital combination (1 through 5, or 1 through 0) in a series of push-buttons to determine the code. It is used to deny access to an individual who is not authorized or cleared for access. It should be considered as an auxiliary lock, not a primary lock, because it does not deny access to an individual who is willing to spend some time working on the mechanism.

Padlocks Padlocks can be either key-actuated or combination. The effectiveness of either depends on their ability to withstand picking or forcing by an intruder who breaks the lock with a heavy object or pries it apart with a bar or cuts the shackle with a hacksaw or with bolt cutters. The standard, easily available keyed padlock should not be used for high-security applications. High-security padlocks, such as those produced by Sargent and Greenleaf (S&G), are the best available. They use the Medeco high-security keyed cylinder and cannot be picked.

Combination padlocks (S&G brand) are the best available for medium-security applications. They have a relocking capability and are of heavy construction and very resistant to cutting, direct force, prying, and manipulation.

Miscellaneous Locking Devices Overhead doors, sliding doors, metal accordion doors, and the like may be locked by various means, and each installation will require close scrutiny. Such doors are usually installed at vehicle entrances and loading docks. If possible, the doors should be locked from the inside only. If the doors are electrically operated, the controls should not be accessible to a potential intruder. The switch should be locked in the "off" position after door is closed and secured at night. When rear or rarely used side doors are used, consider heavy iron or wooden bars installed across the inside of the door and jamb. After the bar is hung in place on appropriate supporting brackets, it is a good practice to drill holes horizontally into the brackets and bar and insert metal pins to hold the bar in place.

Basic Lock Parts

Locking Bolt A metal bar (bolt) that, when actuated, is projected (thrown) either horizontally or vertically into a retaining member (such as a strike-plate) to prevent the door from moving or operating. It also includes the mechanism for projecting the bolt or latch.

Key Mechanism or Cylinder The cylindrical subassembly of the lock, containing the cylinder core, tumbler mechanism, and keyway.

Strike A metal plate or other configuration attached to or mortised into a door jamb to receive and hold a projected latchbolt and/or deadbolt assembly in order to secure the door to the frame.

Locking Bolt Types

Plain Latchbolt A beveled spring-action bolt that is automatically retracted upon contact with the lip of the strike. This bolt is easily opened by even the most uninitiated burglar.

Deadlocking Latchbolt Similar to the plain latch, this typically contains a plunger that automatically locks the projected latchbolt against end pressure when it is depressed. This is considered an average locking bolt. It will deter or delay, but it will not prevent access to an individual who is determined.

Deadbolt A locking bolt that does not have a spring action (as opposed to a latch bolt). It *must* be actuated by a key and/or a thumb turn. When a deadbolt is projected, it locks against its return by end pressure. It is considered the best bolt to use against unauthorized entry.

Deadbolts can be either:

- Horizontal, with movement to the left or right.
- Pivoting, used in narrow stile-type locks, with the projected movement involving a pivoting action.
- Interlocking, employing either vertical, horizontal, pivoting, or a combination of these actions. The bolt and strike are configured to interlock the bolt with the strike to prevent spreading.

Keyed Lock Mechanisms

Pin Tumbler A locking cylinder employing pins to retain a rotating core or plug in one position until the correct key is inserted. It can have five to seven pins; the higher the number of pins, the better the security level.

Wafer or Disk Tumbler A double-action, spring-loaded, flat plate-like mechanism designed to move up and down in slots that move through the diameter of the cylinder plug. Such locks can be easily picked, and the lock construction is such that the entire lock can easily be damaged or destroyed to effect entry.

Warded A minimum-security lock containing internal obstacles (wards) that are intended to block the entrance and/or rotation of all but the correct key. The key may actually engage the bolt to retract or project it. It is more commonly referred to as a "skeleton key" lock. This is the lowest possible form of security protection, almost equivalent to none at all, because it can be easily forced or opened with a skeleton key purchased almost anywhere.

High-Security Cylinder A key mechanism employing a unique, high-tolerance key and tumbler mechanism designed to retard and defeat picking attempts, drilling, and unauthorized key duplication penetration attempts. Provides the best security against attempted picking of the lock.

Removable Core Cylinders A control key is required to remove a specially designed cylinder core containing the combinated tumblers from the cylinder. A single control key can remove all cores within the system. A different core is then reinserted into the cylinder. This type of cylinder is commonly used where one or more levels of master keys are present for standard or medium-security applications. When used in a high-security application, it must contain a high-security cylinder. Most companies that sell removable core cylinders will put in a high-security cylinder if requested.

Note that some of these lock types are combined, such as a high security, removable core cylinder lock.

Lock Strikes

Plate Strike The plate is normally made from flat steel and mortised into the door frame, being secured with two to four attachment screws. It is usually used in conjunction with certain rim cylinder locking devices. The majority of these strikes have short screws that will penetrate the frame up to ¾ inch; all such screws should be replaced with screws at least 1 inch in length.

Mortised Box Strike A flat metal plate backed up with a light-gauge metal box. The entire unit is then mortised flush into the door frame. The metal box provides a better finished appearance but provides no additional security. It is almost always secured with only two screws.

Rim Box Strike An enclosed strike used in conjunction with many of the currently available rim locking devices. A portion of the strike that receives the bolt extends beyond and outward from the door case molding.

Interlocking Strike Consists of a strike that is used in conjunction with an interlocking bolt mechanism.

Open Back Strike Employed in double door exit (panic) situations in which *both* doors are required (usually by fire codes) to be operated independently. The door is also secured with a vertical rod exit device on the inactive side. The other door is secured to the inactive door by a mortise lock equipped with a latchbolt. The open back of the strike allows the vertical, rod-equipped door to open and pass the projected latchbolt of the active door.

Roller Strike Employed only in conjunction with a rim-mounted latchbolt. This is found on electrically controlled doors in *some* instances. The bolt is secured to the door frame by a small keeper constructed of a single roller mounted on a small plate.

Electric Strike Uses electrical energy to release or pass a standard door-mounted latchbolt.

High-Security Reinforced Strike Employs a heavier grade of steel, a unique design, increased length, and a longer offset of attaching screws for an equal load distribution.

Dustproof Strike Installed in the opening threshold or in the floor to receive a flushbolt; equipped with a spring-loaded follower to cover the recess and keep out dirt when the flushbolt is not in place.

As in the lock cylinder types, the strikes can have a combination of features. However, the name of the lock is not usually a combination of both types.

Key Control

A good lock and key control and issuing system is of primary importance in the safeguarding of property, equipment, materials, and information. This system should extend even to the control of security container combinations.

For effective control, there must be accurate record keeping, periodic physical inspections, and inventories. The main principles of such a system include:

1. Combinations or keys should be accessible only to persons who require them.
2. Combinations to security containers, combination padlocks, and keyed padlocks for security containers or other equipment (doors, windows, access holes, and so forth) should be changed at least once every twelve months. They should also be changed at other appropriate times, including:
    - When a key or combination is lost, temporarily misplaced, or when a compromise of the key or combination is suspected.
    - Upon the discharge, suspension, or reassignment of any individual who has knowledge of the combination.
    - When a security container is put into service.

The rotation of keyed padlocks may be required or even dictated by the local situation or other policies. This is a recommended practice for all situations.

In the selection of combination numbers, avoid multiples or simple ascending or descending arithmetical series. Also, avoid numbers from an individual's telephone number, address, age, social security number, or date of birth. For the greatest security, combinations must be random.

Combination padlocks with a fixed combination may be used with bar lock containers as supplementary locking devices. An adequate supply must be maintained to permit a frequent interchange of locks among users. This type of lock does not provide any true level of protection unless it is used in large numbers over extensive areas; this practice permits a successful interchange without compromise. Fixed-combination locks should never be used to protect sensitive or high-value materials, but only for general supplies, low-value equipment, and the like.

Records containing the combinations should be given the same security protection category as the highest level of material authorized for storage in the container which the lock will secure.

The use of keyed locks must be based on the same general principle that is applied to combinations. Issuance of keys must be kept to a minimum and retained under constant key control supervision. If feasible, the key system should be under direct control and administration of the facility security office.

The following measures are recommended for the control of keys to supply, warehouses, sheds, and other structures containing classified information or pilferable items and materials of value:

1. Keys should be stored in a locked container when not in use.
2. Access lists for persons authorized to withdraw keys to classified or other sensitive or valuable facilities should be maintained in a key storage container.
3. Keys should be issued for personal retention or removal from the facility.
4. Key containers should be checked at the end of each shift or day, and all keys must be accounted for.

Key Control Records

A key control record system should be maintained on all key systems at the facility. Accountability can be maintained by records, key cards, and key control registers. Each record must include, at a minimum, the following information:

1. The total number of keys and blanks in the system.
2. The total number of keys by each keyway in use. Keyways should be coded.
3. The number of keys issued.
4. The number of keys on hand.
5. The number of blanks on hand for each keyway code.
6. Persons to whom keys have been issued for personal retention. Inventories of key systems should be conducted annually.

Requests for the issuance of new, duplicate, or replacement keys must be approved or monitored by the official responsible for key control.

There should be a key depository at the installation where the keys are secured during nonoperating hours. Supervisors should sign a register for keys at the beginning of each working day and turn in all keys at the end of the working day. Security personnel should check the key board and register to ensure that all keys are properly accounted for. Key control systems should be devised to provide the required degree of security with only minimum impairment of the facility operations.

Permanent retention of keys by employees should be severely restricted or the keys should be limited to general office work areas. The supervisor should maintain the key for storage or supply buildings and such areas within.

The basic requirements of a good key control system include:

1. High-security pin tumbler cylinder locks will normally be specified and installed.
2. Key control systems will be developed to insure against usable keys being left in the possession of a contractor after completion of work or in the possession of other unauthorized personnel. Specifically, the system of checks and balances consists of key control registers and random checks of key holders. Such assurance is normally achieved by using locks with restricted keyways and issuing new keys on key blank stock that is not readily available to commercial key cutters in the geographical area of the facility.
3. Master keying should be prohibited. Larger facilities will have a master key system for convenience regarding the number of keys employees are required to carry. In such instances, master keying should not be used for sensitive, high-value, research and development, cash flow, or other areas that may involve some level of sensitivity. This includes senior officials' office areas. When pin tumbler locking devices are master keyed, the use of several shorter pins to facilitate two or more acceptable pin positions reduces the security afforded by the use of a maximum number of pins in a nonmaster keyed lock.

One or more mushroom-type pins or a variation of this type of pin can be used in each such lock at facilities where high security is an integral part of the facility security plan. Also, individual pins should not be segmented into more than two for each pin chamber on those locks that are used to secure more sensitive materials and information.

4. All locks and keys in a master keyed system should be numbered with an *unrelated* numbering system. "DO NOT REPRODUCE," "DO NOT DUPLICATE," or "DND" will be stamped on all master, higher level, and other control keys. If feasible, consider a "DND" stamp on all company keys. Note that by having a controlled keyway, you also severely limit an individual's ability to duplicate the key at a commercial key cutting establishment.

Key Control Officer

An individual within the security office should be designated as a key control officer. The position should not be left to building maintenance, engineering, or administration. This individual will be responsible for ensuring that an adequate supply of keyed padlocks, combination padlocks, door locks, and other locking devices required by the facility are on hand or are available through a local source; the handling of keys; records maintenance; investigation of lost keys; inventories and inspections; custody of any master keys and control keys, if applicable; regulations and local policy concerning locks and keys for the installation; maintenance and operation of the facility key depository; and the overall supervision of the key program for the facility.

The key control officer should be maintaining (at a minimum) a permanent record of the following information:

1. Locks by each number, showing:
   - Location of each lock;
   - Key combination (pin lengths and positions); and
   - Date of last key change.
2. Keys by number, showing:
   - Location of each key;
   - Type and key combination of each key; and
   - Record of all keys not accounted for.

The key control officer is also responsible for the procurement of all locking devices and associated keys and/or key blanks. Based on facility requirements, he or she should coordinate their procurement, and keep informed of improved locks, keys, and key systems that may be applicable and advantageous to the facility.

A key control register is maintained at some central point within the facility. Keys will be signed out to authorized personnel, as needed. The register contains the key identification number, date and time of key issuance, printed name and signature of recipient, the initials of the person issuing the key, the date and hour the key is returned, and the initials of the person receiving the returned key. When not in use, the key control register will be kept in a locked container, usually the same place where the keys are maintained.

In the recent past computerized key control has become available. Administrative details can be easily entered into a computer file and updated. There would also be a simple notebook in which individuals would actually sign for each key received. Such information, of course, would be put into the computer to maintain the record electronically. At regular intervals a hard copy of the data should be printed to provide a record of the current key control status.

The key control manager should maintain a backup computer disk, keeping it separate from the initial day-to-day disk in case of an accident or emergency that would damage the primary disk. Never enter or maintain your key codes in your computer hard drive.

A locked container or key repository can be an ordinary safe or security container, but it can also be a filing cabinet or a key depository made of at least 26-gauge steel, equipped with a tumbler locking device, permanently affixed to a wall. When such a unit is used, it should be located within a more secure area, not in a typical office where unauthorized personnel could gain access.

For facilities that maintain a twenty-four-hour-a-day guard system, a limited number of more generally used keys may be kept at the central guard post and signed out by individuals. This kind of setup is necessary and vital in an emergency or in unusual situations. All such keys will be signed out under the same conditions that are in effect when receiving a key from the key control officer. No keys should be maintained at the guard post that would allow entrance into sensitive, restricted, controlled, or high-value areas *unless* the keys are locked up and are under the direct and personal control of the guard shift supervisor.

All doors used for controlled access—or for areas that contain valuable or highly pilferable items—must be locked with a medium- or high-security locking device. A low level security lock is too much of a temptation; it is simply too easy to defeat by even the uninitiated. The locking device on the most secure door must be a high-security padlock and a high-security hasp. The secondary padlock, mortise lock, or rim deadlocks must be used to secure the other door or double door requirements at the facility. Mortise locks and rim deadlocks should meet the following specifications:

- Be key operated.
- Have a deadbolt throw of at least 1 inch, although 1 ½ inches is preferred.
- Be of a double cylinder design, where facility policy applies.
- Cylinders are to have five pin tumblers, two of which are to be of the mushroom or spool-type drive pin design for a master keyed lock.
- Have 10,000 key changes possible.
- No master keying of a lock will be permitted without approval of security key control officer.
- If the locking bolt is visible when locked, it should contain a hardened, saw-resistant insert or should be made of steel.

Vehicle garages or vehicle storage area gates will be secured with an approved secondary padlock.

Doors that cannot be secured from the inside with locking bars or deadbolt locks will be secured on the inside with a secondary padlock and hasp. The hasp will have at least 1 ½-inch screws. The hasp should be made of steel, never lightweight aluminum.

Door Standards

In general, an intruder's attack on a door centers around one point—the locking device—but should this fail, he will try another method. Penetration may result from the failure of any number of different door system components, namely, hinges, glass in the door, frame spreading, or prying to bypass the locking bolt.

A door system, as a whole, is a unit composed of a group of parts that make up a closure for a passageway or a wall, including the lock, bolt, and strike-plate; the actual door; frame; and hinges.

Door Types and Materials Doors fall into three basic categories: flush, panel, and stile and rail. Flush doors can be characterized as plain, that is, having no face panels, louvers, or recesses. They can be hollow core, solid core, particle board, or of glued block construction. Doors are made of wood, metal, or glass.

A panel door is the same as a flush door, except that it has one or more pieces of glass or louverd panels installed. The stile and rail door has wood and/or glass panels that are separated by independent vertical stiles and horizontal members.

Door material can be wood, steel, aluminum, or speciality types (combination, custom-made, entirely glass and/or ornamental in nature). Door frames are made of wood, steel, or aluminum. Wood frames should be made of a medium or hard wood. They can be of a standard milling or created as separate pieces and then installed to produce the final product. Steel frames are constructed of extruded steel. A stop may be applied to or formed with the frame during the extrusion process. The entire frame usually comes ready to install with mitered and welded corners. The door and frame can come in several separate sections and be field-installed and the corners welded during installation.

Aluminum frames—also made through an extrusion process from heavy aluminum tubing—are now used extensively in conjunction with glass front doors. The stops for aluminum doors may be formed but more often are of an applied type.

Hinge Assemblies Hinge assemblies or leafs have one part that is attached to the door and another part that is attached to the frame. They allow the door to pivot or swing either inward or outward, from the closed to the open position and then back to the closed position.

Hinges come in several common varieties:

- Removable pin—This is a standard household hinge with a pin holding the two leafs together. The pin can be easily removed.
- Fixed (nonremovable) pin—A variation of the standard hinge that has a pin that cannot be removed once it is installed.
- Offset pivoting hinge—This is mounted between the center and hinge stile at the top and bottom of the door. It is designed to pivot the door open rather than hinge it at the door side to the frame.

- Integral Hinge—This is a hinge located on the hinge stile at the top and bottom of the door and is installed as a part of the door.
- Continuous hinge—This is used primarily on aluminum door systems; very rarely seen on any other type of door. It has no hinge knuckles or pins. Instead, it is affixed along the entire length of both leafs in a series of gearlike projections, which are interlocking.

Door Frame Protective Measures Table 7.1 illustrates accepted measures that can be used to protect a door from having its frame spread. These methods allow the door to properly function, and they give the door system a much stronger frame and also tie the frame to the supporting wall to which it is attached.

TABLE 7.1 Door Hardware Maintenance

The proper operation of door hardware means you should have a program in place which consists of the following:
1. All door system components checked at least quarterly or with seasonal changes. The expansion and contraction of the door and/or frame can cause problems.
2. Clean dirt, grit, and other materials from the threshold, and from around mounted strikes.
3. Remove any foreign material found on latchbolts.
4. Lubricate the locking mechanism, following the lock manufacturer's recommendations.
5. Adjust the strike to ensure proper alignment with the latchbolt.
6. Adjust the door closure pressure, if used on the door.
7. Check the door alignment. It should not bind, but move freely.
8. Check the door closer, if used, for leaking seals. If the fluid type is used, check the level.
9. Check hinge pins; they should not be loose.
10. Tighten or replace any loose or damaged screws. This includes those associated with the lock cylinder, knobs, strikes, lock faceplates, hinges, door closers and other hardware that may be affixed to the door assembly.
11. Number the doors in some manner so that lock or door problems can be more accurately reported (only for larger facilities with numerous doors).

Windows

Windows are the second most common point of entry into a building. All windows should be considered vulnerable when:

1. They are less than eighteen feet from the ground or are easily accessible from the building exterior.
2. They have an opening of at least ninety-six square inches or more, or one side of the opening has at least eight inches vertically and fifteen inches horizontally.
3. They are forty inches or less from a locking device that could be activated by an individual reaching through a window-like opening, either by hand or with a hand-held object.

Windows can be of the following types:

- *A double-hung window* is the most common and is standard in almost all homes and office buildings. The window has a vertically sliding two-sash unit, which can be opened from the top or bottom. It is commonly secured with a single rotary sash holder. Frames can be wood or metal.
- *The casement window* has a sash hinged on the side. It is usually opened inward or outward with a hand crank located at the bottom of the window frame.
- *A horizontally sliding window* operates on the same principle as the sliding door.
- *An awning window* has sash hinges at the top and opens outward. It may also employ a hand-crank operation and, if so, it can operate inward or outward.
- *The hopper* is basically the same as an awning window, but it is hinged at the bottom and only opens inward.
- *The jalousie window* consists of small, frameless glass panels that operate simultaneously with a hand crank located along the side. These types of windows are mainly used on homes in warmer climates, but are frequently found in smaller or older private, industrial buildings and government facilities that have been leased. Small metal tabs fasten the glass and can be easily bent outward and the glass removed to allow access. Jalousie windows should *never* be located anywhere along the perimeter of a facility building.

8
Facility Access Control

Facility access control systems and devices are those controls that assure that individuals and vehicles entering the facility grounds are authorized. Access control systems have different specifics within the type of program developed for each facility and are based on an evaluated need.

Some of the first questions asked are

- Do you really need an access control system? If so, how simple or complex must it be?
- In what circumstances would you need the system?
- Is the proposed system cost effective?
- If an off-the-shelf type system is devised, will it meet all the facility needs and requirements, or will you have to revise it?

In terms of the need, some further questions arise. These are the most *basic* questions that must be answered to determine if the facility really requires an access control system:

- Does the facility have areas that need to be secure, that are within the facility barrier confines? Do these areas generate a

reasonable amount of traffic flow during the beginning and end of each day?
- Does the facility have a requirement for various levels of access and denial to various areas?
- Is a record required of who entered and departed from various facility areas at certain times?
- Is there parking, fuel or refueling storage points, or other areas outside the facility buildings that require limited access only for authorized personnel, while denying access to all others?
- Does the facility have a need to restrict employee and visitor access to the facility, to individual buildings within the facility, or to specific facility rooms or areas at certain times of day, night, and/or weekend?

You should be looking at the access control system circumstances, keeping in mind that the following have been partially answered from the above.

- Would a hard-copy record reflect all entries into restricted or controlled areas and would this be beneficial and sufficient for your record keeping, particularly in the case of an internal investigation?
- Are there sufficient personnel to watch and supervise all entry/exit areas into restricted or controlled areas of the facility? This is in addition to any personnel used at main gates for facility access.
- Is it advantageous to avoid issuing or signing out keys in storage, fuel, supply, sensitive, restricted, and controlled areas by increasing guard personnel at these location entrances?
- Within the facility is there a high incidence of the wrong people being in the wrong place at the wrong time?

Cost effectiveness is the last area of concern:

- Are security personnel currently placed at access points to controlled or restricted areas?

- Is it taking longer amounts of time to get authorized personnel (plus appropriate visitors, as necessary) into controlled or restricted areas?

If most of the above questions can be answered with a YES, then you should have an access control system of some type. Such a system need not be expensive, it could be a simple one that produces a basic picture ID card for each employee. Some measures may actually cut costs. If you have or are considering a CCTV system for surveillance of controlled areas, security guard personnel that would normally control access to some areas can be reduced. From a budgeting standpoint, payroll savings for a larger facility will probably cover the costs of purchasing or developing the appropriate system in-house.

In determining what system to use, request proposals from several sources. Specific dollar figures should be included in the proposals. Compare proposals against each other *and* with your system specifications and requirements. Look at training, backup support, and response to system malfunctions and then make the decision.

These considerations are helpful when evaluating a facility that may not have a system and may be considering one, and also when looking at a facility that has an access system to see if it meets all necessary facility requirements.

PERSONNEL ACCESS CONTROL SYSTEMS

Manual Access Control

A manual system relies more on the human factor because it involves visual recognition by security force personnel prior to authorizing access. For a manual access control system, a facility/agency identification badge should be used. The badge should be designed, developed, printed, and controlled by the security office of the facility. Likewise, the issuing requirements for full-time, part-time, and maintenance personnel on-site, in addition to badges for visitors, VIPs, and vendors must be developed and implemented. Such procedures include requests for badges to be issued, requirements for various types of badges, and the duration of authorization. The processing, administrative paperwork, badge recall, accountability, and the destruction of imperfect and outdated badges must also be taken into consideration. The facility security office must maintain a quantity of the various types of badge blanks that are used in the conducting of facility business. Finally, any badges that are prepared (whether or not issued) *must* be sequentially numbered for control purposes.

Where government contracts and classified matter are included, background checks and appropriate clearances must be on file at the facility prior to the issuance of a permanent or a temporary identification badge.

Electronic Access Control

The electronic access control system is one that uses a basic signal tied to a processing unit to determine whether or not a particular access card (the ID badge) is authorized for entrance into that particular facility or portion thereof.

Control system types to consider (not applicable to personal recognition systems as used with a basic identification badge system):

1. Local nonprogrammable—An off-line system that recognizes one, two, or three group codes selected within the card reading unit. This has limited applicability within standard government facilities.
2. Local programmable—An off-line system that can accommodate a number of cardholders (usually 1,000 or less) and may be individually programmed to allow access into and out of the system. This has several access levels depending upon the design of the system functions.
3. Remotely programmable—An on-line system that provides for programming of a variety of functional levels of access and denial, and also stores a large quantity of cardholder identification information.

Access control systems have three functional elements that must be present for the system to operate properly:

1. Encoding element—Identification information is entered into the system and requests for actions are initiated. This includes both the code and the methods employed for reading the code.

2. Control element—The data received from the encoding system is routed, processed, stored, compared, and/or otherwise used to make a decision (access or denial).
3. Action element—This is the event that results from the utilization of the data from the control element.

Any electronic access control system will have one of four types of encoding methods for personnel to access the system and have their card read.

1. Personal recognition—This involves person-to-person contact at the encoding element and requires that the person granting access have the authority and ability to determine the action.
2. Combination—The access unit employs a group of numbered or lettered buttons that, when depressed in the proper sequence, allow the access action to take place.
3. Personal characteristic—The encoding method utilizes some personal and some physical attributes of the individual wishing to enter.
4. Card System—Utilizes various encoding techniques to magnetically, optically, or electromagnetically store information on a cardlike material; only that data is used in a decision to grant or deny access.

The fourth type of electronic access control is used in 90 percent or more of the currently available systems. The first three types are for earlier systems.

From a security point of view, the control of personnel entry is of prime importance. If entry cannot be controlled, then there is no method for disallowing unauthorized personnel.

The security office must develop applicable procedures for the positive identification and control of employees. Sample copies, marked as samples and controlled, must be available for security force personnel to determine the authenticity of a badge, if necessary.

The identification badge for the system is usually factory encoded with certain basic information. Additional controls are built within the card reader system, namely, to grant or deny access to a specific area. The changes are not on the badge, but within the system itself.

Badge blanks must be strictly controlled within the security office and only personnel authorized to put information on a badge, process the badge, and issue the badge to personnel should have access to the badge blanks. All badge blanks must be sequentially numbered prior to or immediately upon receipt of the badges. The accountability for badge blanks is a major security concern that cannot be overlooked.

Specific requirements and procedures must be developed for lost, damaged, or stolen badges. Consider limiting the amount of time before a new badge can be issued in the case of a lost badge.

The frequency of visual checks of the badges should be built into the security system. Sometimes, a random check or a 100 percent check over a two- or three-month period will suffice. If a badge is deliberately stolen, an immediate 100 percent check for several days may be necessary.

Receipts must be prepared for all badges issued and the signatured receipt filed. The system processor should enter the appropriate data regarding individuals' specific badge number(s) and type(s) of access.

Badges should be coded—by number, color, or a vertical or horizontal stripe—to indicate further controls within the facility.

Facilities that have large numbers of visitors will want specifically designed badges, numbered and controlled, for visitors. The badge design, color, and other pertinent data will indicate whether or not an escort is required and the level of access and/or clearance for the individual temporarily retaining the badge.

BADGE EXCHANGE SYSTEMS

Badges for VIPs and vendors must be considered. The design and color should be significantly different from other badges. Some facilities may have a badge exchange system whereby one badge is exchanged for another to gain access to tightly controlled or restricted areas. The badge exchange may or may not be one that involves an electronic access control system (EAS); current technology has made such systems functional at numerous facilities across the nation.

The badges used in an exchange system should be basically the same in terms of format, identification information, and the user's picture, but one or more unique items should be different. For example, the basic badge is printed in blue-black ink on a cream white paper.

The exchange badge is identical except the ink is red and the paper is light grey.

Badge control numbers should be the same on the initial and the exchanged badge. In such systems, the badge should be effective no longer than twelve months, and then a new badge would be required. Because of the sensitivity of areas requiring a badge exchange system, badges for all personnel should be changed every third or fourth year, with a completely different design, color scheme, background imprint, and numbering.

Vendors, visiting repairmen, and janitorial personnel who regularly enter and/or work in large facilities often become very familiar with the operations and permanent party personnel and are able to locate storage areas of valuable supplies and equipment, money, and sensitive areas. They are in the unique position of being able to get to various areas of a facility without arousing suspicion. They are then in an excellent position to "case" the building area, contents, alarm considerations, and guard controls. As outsiders they obtain an insider viewpoint.

All such personnel, unless permanently on-site or known to maintain a high-security clearance level, should be escorted in and out of the facility and be under the watchful eye of an employee or security escort at all times. Their movement should be limited, so that they use the most direct route possible to and from their work point to their access/exit point.

FACILITY ADMITTANCE REQUIREMENTS

All facilities should have set procedures and requirements to allow entry. Such procedures and requirements may be posted or set forth in specific regulations that are distributed to all personnel. The admittance to any facility must include the following within specific requirements:

1. A site-issued/agency-issued identification pass for entry into and past any security checkpoint.
2. The general work hours and security hours must be established. They must be in writing. Generally, standard work hours are Monday to Friday within the time period from 0600 to 1800 hours. With flextime in effect for many government

and private concerns, personnel work a certain number of "core" hours—hours when everyone must be on-site—and then optional hours, which allow personnel to start earlier or later each day, while ensuring that they work a minimal number of hours.
3. The security hours are designated as those in which authorized building pass identification cards are required to gain admittance and remain within the specified limits of a facility. The security hours may apply to only a portion of the facility or the entire facility. When hours apply only to a portion of the facility, other portions that are unaffected by security hours may require only general identification (a facility auto sticker, for example) for entry. For those areas within the security hours control, a sign-in/sign-out log or other system should be used to permanently record names of individuals who entered and exited the site. When an entry control system is in use, it should be able to extract data from a card that is used to provide a daily printout for later review.
4. Admittance to security areas within a facility should be in accordance with specific policy and procedures prescribed to allow personnel working within that area to be given entrance.
5. Persons with official business who lack appropriate facility-issued passes must be escorted unless there is a valid security clearance on file and/or a permanent or temporary pass is authorized. Such passes must have a cutoff time after which they are no longer valid. Where a security guard is stationed on a twenty-four-hour-a-day basis, the badge is signed out and returned when the visitor leaves the facility.

Vehicular Access and Controls

Vehicle access and control procedures vary from very little to somewhat stringent, depending on the facility size and location. Overall, the controls are the same for both employee and nonemployee vehicles, but the depth of the checking may differ, according to the facility policy and location.

In larger facilities, parking may be located outside any controlled access area. It then requires some form of vehicle pass or sticker for entry and authorized parking. Without such a pass or sticker, a person would have to stop, sign in, and be directed to a visitor space by the security guard on duty. Parking lots lacking a security guard lose control over visitor access. Many controls are not instituted until the visitor reaches the next line of protection for the facility; in some cases, that line is inside a building.

Normally, facility employees who need to obtain parking permits and decals for their vehicles will be processed through the personnel and/or security office. Nonemployee personnel, including part-time workers, tradesmen, visitors, and others will have to go through this process each time they enter the facility. Such personnel would not have permanent parking privileges within the facility perimeter.

In the case of a facility built upon a larger installation, the installation parking policy and rules will apply to all. In such cases, the individual facility would not take any active part in parking access and control, *unless* the facility parking area is inside its own facility perimeter. In such case, there may or may not be an internal permit for access within the facility grounds. This would have to be determined at the local level.

While it is expected that all facilities will have areas set aside for employees and transient visitors to park, this is not a mandatory rule. Facilities within commercial areas of cities may have employees park on the street or in commercial parking lots. Some facilities, however, may have a portion of their building set aside for employee parking, usually in the basement area.

The area(s) where employees park may have disadvantages, specifically the theft of portable office equipment and supplies. For this reason, security inspections on a random basis at exits, in addition to the requirement for property passes, will prove extremely useful.

Whenever possible, parking should be maintained away from entranceways or the sides of a building if there are numerous opening windows.

An alternative, and one that is recommended, is that all employees, including VIPs, park a short distance away from the building. If possible, locate parking areas outside the security fence or wall. This can reduce the potential for facility damage from gasoline fires in vehicle tanks, in addition to minimizing the hazards of explosives and/or incendiary devices concealed within vehicles.

This alternative is not reasonable when the facility is large, but if vehicles must be inside the perimeter line, they should be parked away from the fence a short distance (at least twenty-five feet) to allow for better perimeter security visibility and away from the building entrances, supply dock, or facilities.

Vehicle parking permits should be sequentially numbered and color-coded to indicate specific parking lot areas. Decals can be used as the authorization for a vehicle to gain access and park, or in addition to the permit. In most cases, the parking permit is placed in the window, while a decal is affixed to the left front/right rear bumpers of the vehicle. Vehicles that have regular access should have a decal for facility access *and* a parking permit authorizing specific area parking.

There must be definite procedures for the registration of all vehicles. Such procedures are to be in writing and promulgated for all employees. Procedures should include complete vehicle identification and registration, including proof of vehicle insurance. Registration should be updated on an annual basis, with no exceptions. An individual with an expired decal/permit should be treated as a visitor until the vehicle registration has been updated.

A vehicle's decal must be such that it cannot be removed from a vehicle without defacing it. This avoids the problem of an unauthorized person removing a decal and placing it on another vehicle for unwarranted purposes. Loss of theft of a decal and/or parking permit must be reported immediately so that it can be canceled and the security guards notified.

Random checks should be made of parking lot areas for vehicles that have been left overnight and also to see if personnel are using authorized parking area spaces.

Visitor parking spaces should be away from regular employee parking areas, but close enough that security force personnel can observe them. They should never be near entrances to the building.

All visitor vehicles need to be signed in to the facility and assigned to a *specific* parking space. A temporary parking permit is issued. This permit must be of a completely different color and/or design than permanent parking permits. When leaving the facility, the permit must be turned in at the same guard post from which it was issued. During evenings, weekends, and holidays visitors should not be allowed to bring vehicles into the facility.

Trade personnel and/or other facility support—including maintenance—should be issued individual permits when entering the facility and directed to specific parking areas or to specific loading dock areas for the delivery or pickup of supplies and equipment. A telephone check by the security guard to a guard at a loading dock or main desk will confirm the time the vehicle arrives and departs. (Note: Sometimes such personnel "tour" the facility in their vehicle when first entering or when departing. This can be a form of "casing" if they are involved in a theft ring or if they are involved in espionage, sabotage, or terrorism. A spot-check to note the time they leave their work area (such as a loading dock) and the time they sign out should be made and any inconsistencies investigated.)

9
Security Containers and Storage Areas

Corporate, sensitive, or classified government material or information should be used, held, or stored only where there are appropriate protective measures in place or under other conditions that are adequate to prevent unauthorized persons from gaining access to it. The exact nature of the security requirements will depend upon a thorough security evaluation of local conditions and circumstances. These conditions must permit the facility to accomplish its essential functions, while still affording selected items of information a reasonable degree of security protection with a minimum degree of calculated risk that these items will be exposed to unauthorized personnel.

Material and information must be stored in approved security containers, vaults, or rooms.

Items having only a monetary value, such as cash, precious metals, jewelry, narcotics, and the like should be stored separately from classified information or other sensitive corporate data. Store these types of items in containers or rooms specifically designated for them. They are *never* intermingled with unlike items.

The most sensitive materials and information are stored in locations that are augmented by the facility security procedures to prevent unauthorized persons from gaining access to the container, vault, or room. The location of the storage equipment within a building or an individual room (which is locked, guarded, and/or alarmed during nonoperating hours) will have to satisfy any minimum security protection requirements determined by a government contract, commercial contract, insurance requirements, and common sense security practices and procedures.

STORAGE EQUIPMENT STANDARDS

Storage equipment standards (to include rooms and vaults) are based upon established and published uniform standards, specifications, and schedules for such containers, vaults, alarm systems, and/or other associated security devices that are suitable for the storage and protection of classified information and material.

For government and government contractor organizations, the General Services Administration (GSA) issues the appropriate standards. Individual government departments or agencies also have their own security policies and procedural requirements. Here, we will use the most commonly held standard security requirements, that of the Department of Defense (DOD) and the DOD regulation 5200.1-R.

Facility heads can also establish additional controls to prevent unauthorized access. Such controls are usually developed and coordinated by or through the facility security office and disseminated under the direction of the facility head.

Government security storage containers will bear a "Test Certification Label" on the locking drawer, attesting to the security capabilities of the container and the lock. On some older cabinets, the label is affixed on the inside of the locking drawer compartment. Cabinets manufactured after February 1962 indicate "General Services Administration Approved Security Container" on the outside of the top drawer.

Substandard storage equipment of facility areas, for example, power shelf files, rooms for open storage, or containers not listed on the approved GSA schedules, *may* be authorized only for the *temporary* storage of material, but they are subject to inspection and approval by the site security office prior to their temporary use. The requirement for additional corrective security measures, namely, access controls, alarm systems, special locks, vault-type doors, special construction, as well as the imposition of certain storage limitations upon the user, are not uncommon.

The GSA-approved changeable combination padlock is intended for use only as an indoor or sheltered area protective measure. It is not intended for use either outdoors or to protect against a determined forced entry. This is the only padlock approved for use with locking bar containers. Because of their vulnerability to force, locking bar containers must be limited to the storage of low-level information (confidential for government areas) unless they are situated in a vault or alarmed area. The padlock used is the Sargent and Greenleaf Model 8077 series.

Site security offices should, upon request, inspect and evaluate any proposed storage area to determine the type, amount, and degree of security required to afford physical protection equal to or greater than that authorized for classified materials. This will include the various authorities with regard to regulatory requirements for the protection of special access materials.

STORAGE OF GOVERNMENT CLASSIFIED MATERIAL AND INFORMATION

Classified material and information that is *not* under the personal control and observation of an authorized person must be guarded and stored in an appropriately locked security container.

Top Secret information will be stored in a security container having a three-position dial combination lock that has been GSA-approved or else stored in a Class A vault or a vault-type room that meets the standards established by the head of an agency or department.

When such containers are maintained in buildings not under U.S. government control, structural enclosures, or other areas, the storage equipment, vault, or vault-type room *must* also be protected by an alarm system or guarded during nonoperating hours.

An alarmed area—provided that the area can afford protection equal to or better than that indicated above—is used for storage of classified material. This ensures that the protecting physical barrier is adequate to prevent the surreptitious removal of the material and also to prevent observation that would result in the material being compromised. The barrier must be such that a forcible entry attempt will give evidence of the attempted entry into the area. Further, the alarm system must provide an immediate notice of alarm to the security force that an attempted entry is being perpetrated.

The security guard force must respond within a time limit of five minutes at most; the typical response time should be two to three minutes. Actual response time will vary depending upon location of the alarm and location of the responding guard. Note that if at least one individual of the security guard force is a "rover" (on patrol throughout the facility), the response time may vary. Some organizations maintain two personnel at the security alarm system so that one of them can respond immediately. Thus, one person maintains the alarm monitor panel continuously while the second responds to the alarm within the allotted timeframe. This also ensures that the panel is never left unattended. If such conditions cannot be met, then top secret storage should neither be permitted nor authorized.

Supplementing the vault with a security container within, assuming the vault is within a substantially constructed and secured building or room and either guarded or alarmed, will satisfy the minimal requirements for the protection of top secret storage.

Guard force personnel need not be cleared for top secret if they are positioned external to the building, vault, or room and are not expected to ever have access. They should, if possible, or by contract guidelines, have at least a secret level security clearance. Inability to meet these standards will mean that the use of augmented security personnel in the area will be required continually to respond to situations that may arise.

Secret and confidential information should be stored in the manner as authorized top secret, a Class B vault, a vault-type room, or secure storage room meeting appropriate standards; until phased out, in a steel filing cabinet having a built-in three-position dial combination lock; or, as a last resort, an existing steel filing cabinet equipped with a steel lock bar, provided it is secured by an approved, changeable combination lock. In this latter instance, the keeper(s) and staple(s) for the locking bar must be secured to the cabinet by welding, rivets, or peened bolts. Other supplementary controls may also be required.

The actual combination lock used for various types of security containers and vaults must be one that meets GSA standards for government and contractor personnel. The GSA standard is an excellent choice for private concerns also. Specifically, the item of concern becomes: what are the interior tumbler wheels made of—metal or a plastic-like material? If the wheels are metal, the security of the container has just been lowered. Metal wheels are subject to x-ray and the combination can be determined within thirty minutes. Nonmetallic wheels cannot be x-rayed and thus provide a higher level of security against attempted entry. Government security containers that hold classified material cannot have metal wheels. This is also a good argument for private industry to upgrade its containers to protect against possible x-ray of the locking mechanism as a means to determine the combination.

Other storage equipment and requirements include:

1. Field safes and one-drawer containers (able to be carried or easily moved by one individual) can be used for the storage of classified information in situations outside the facility and for the transport of classified information. These containers must be securely fastened or guarded to prevent theft.
2. Map and plan file containers are for the storage of odd-sized items such as computer cards, maps, and charts.
3. Storage areas for bulky materials, other than top secret, must have access openings secured by either changeable combination padlock or a key-operated padlock with a high-security cylinder. The key-operated padlock should have either an exposed or shrouded shackle to protect against forced entry attack upon the lock itself.

When combination padlocks are used, their keys shall be controlled as classified information with classification equal to that of the information being protected (they do not circulate and are not retained by individuals on their person, but are maintained within a security container or other continuous level of protection); a key and lock custodian must be appointed to ensure the proper custody and handling of keys and locks; a key and lock register must be maintained to identify keys for each lock, their current location, and custody; all locks and keys must be audited each month; an inventory of keys with the change of custodian responsible for container and applicable keys; keys cannot be removed from the premises; keys and spare locks must be protected in a secure container; no master keying is authorized; and locks shall be changed or rotated annually and must be replaced upon the loss or compromise of a key.

Hasps that enclose the padlock should be used in high-security applications because they will enhance security by providing the lock a greater level of protection against forcible attack. A chain, when required, should be used with an exposed shackle padlock. Use ⅜-inch tool-resistant, case-hardened security chain or ⅜-inch trade size, grade 80 alloy steel chain.

Additional security safeguards to be applied under these types of storage conditions are detailed in this chapter.

Please remember that perfect or absolute security is always a formidable goal, but a state of absolute security can never be attained. No object is so well protected that it cannot be stolen, damaged, destroyed, or observed, given time and expertise. The purpose, therefore, of physical security standards and requirements is to make access *so difficult* that an intruder will hesitate to attempt penetration or to provide for the intruder's apprehension should the intruder continue to attempt unauthorized entry.

PROCUREMENT AND PHASE-IN OF EQUIPMENT

Any time a security container is requested, facility security personnel must survey current on-hand security containers to determine if it is (or is not) feasible to use available equipment or to retire, return, declassify, or destroy enough records on hand to make the needed storage space available.

New containers shall be procured from those that meet the requirements and are approved. Requests for a new container must specify that it have a three-position, key or hand changeable combination lock installed and that the container come with castors (wheels for ease of movement) already mounted.

Under no circumstances should locking bar containers be fabricated by either converting older steel files or newly procured files to circumvent existing policies and procedures. As feasible and possible, all such containers will be phased out of use for protecting classified material and their storage requirements will be downgraded to other applicable levels for the storage of general office files, materials, and such that are not of a classified nature.

When new containers are procured, ensure that the combinations are changed only by authorized contractor company personnel or by the security office. Individuals should not change any combinations themselves.

CONTAINER REPAIR

Repair of security containers usually calls for the neutralization of lockouts or the repair of any damage that may affect the security integrity of a container. Such neutralization and repair will be accomplished only by those individuals so authorized, who are cleared or continuously escorted while so engaged. In many instances, the actual repair of the container will be effected by a contracted company. At such times, the damaged container and/or container drawer may be removed from the premises by the contractor for repair and then returned to the facility.

A container is considered to have been restored to its original state of security integrity when all damaged or altered parts are replaced. It is also considered restored when a container has been drilled immediately adjacent to or through the container dial ring to neutralize the lockout, the replacement lock is equal to the original container equipment, and the drilled hole is repaired with a tapered, hardened steel pin or a steel dowel, drill bit, or bearing with a diameter slightly larger than the hole. This pin or rod is of such a length that when driven into the hole, there remains at each end a recess greater than $1/8$ inch and less than $3/16$ inch deep to permit the acceptance of substantial welds, and the rod is welded both on the inside and outside surfaces. The outside of the drawer head must then be puttied, sanded, and repainted so that no visible evidence of the hole or its repair remains on the outer surface.

Containers that have been drilled in a location and/or repaired in the manner described above are acceptable. Those that have had a hole drilled in another location, cut, burned, or sliced to gain entry, or repaired in a manner *other* than as specified will not be considered to have been restored to their original integrity. The test certification label on the locking drawer and the "GSA-approved security container" label, if any, must be removed immediately from such containers.

If a damaged security container is repaired with welds, rivets, or bolts that cannot be removed and is replaced without leaving evidence of entry, the security container is limited to the storage of secret and confidential classified information.

When any damage is repaired using methods other than those described, the use of the security container will be limited to unclassified information. A notice to this effect should be affixed to or marked on the front of the security container in such a manner that its removal would be difficult and noticeable. If any combination to the container is maintained in the security office files, this must be indicated on it. Also, the facility listing of security containers should indicate that the container is not authorized for classified storage.

Security containers are usually serviceable for approximately twenty-five years if they have been properly serviced and maintained. Lock or locking bolt linkage malfunctions requiring the neutralization of the container will shorten their life span.

Table 9.1 Signs of trouble that indicate a security container may or does have problems.

1. A dial that is unusually loose (to include in and out play) or is difficult to turn.
2. Any movement of the dial ring (determined by applying a twist to the dial ring to detect this).
3. Difficulty in dialing the combination or opening of the container. Some examples include:
   - The need to dial the combination more than once when human error is not at fault.
   - The need to dial numbers slightly above or below the correct number of the combination.
4. Difficulty with the control drawer or other drawers, in such areas as:
   - Drawers rubbing against the container walls. The container may not be level or the tracks (or cradles) may not be aligned properly.
   - Problems with opening or closing drawers because the tracks (or cradles) need lubricant, material is jammed behind the drawer, or the internal locking mechanism is tripped.
5. Difficulties in locking the control drawer, such as:
   - The drawer handle or latch will not return to the locking position when the drawer is closed.
   - On Sargent and Greenleaf locking devices, the butterfly in the center of the dial will not turn after the control drawer is shut and the dial has been turned to "0."
   - The locking bolts move roughly, slip, or drag, or the linkage is burred or deformed.

Users should be alert for early warning symptoms; when they are detected, the appropriate support offices should be contacted. Personnel should never use force to try to correct a problem. Critically needed material should not be stored in containers that exhibit any of these symptoms, because they are not dependable and the user may be "locked out". Table 9.1 highlights the most common problems.

CONTAINER CLASSIFICATIONS AND LOCK GROUP TYPES

For the government, Executive Order 10101, as amended, established the requirement that whenever new security container equipment is procured it should, to the maximum extent feasible and practicable, also be of the type designated as approved for storage of classified material.

Each approved container, including vault doors, will have a label attached to the front (Table 9.2).

Before 1993, the lettering for the identification label was in black; after that date, the lettering was changed to red.

In addition to the label, the control drawer of security containers and the inside of safe and vault doors will bear a test certification label, as shown in Table 9.3, which will indicate the class of the container and how long the container will resist forced and surreptitious entry, radiological attack, and manipulation of the locking device.

GSA-approved security containers also have an identification label, as illustrated in Table 9.4.

Containers that are used for the storage of government classified information should have all three of these labels. The test certification and identification labels are always located on the side of the control drawer.

Table 9.2 Typical identification label found on a government security container

GENERAL SERVICES ADMINISTRATION
APPROVED
SECURITY CONTAINER
(MANUFACTURER'S NAME)

Table 9.3 Sample GSA security container Test Certification Label

THIS IS A U.S. GOVERNMENT CLASS 6 CABINET WHICH HAS BEEN APPROVED BY GSA UNDER FEDERAL SPECIFICATION AA-F-358G. IT AFFORDS THE FOLLOWING PROTECTION:
- 20 MAN-HOURS AGAINST SURREPTITIOUS ENTRY
- 30 MAN-MINUTES AGAINST COVERT ENTRY
- 0 MAN-MINUTES AGAINST FORCED ENTRY

Contractor facilities, and others throughout private industry, can also use the GSA standards for the protection of company grade materials and are required to use such standards when working on U.S. government contracts with various departments and agencies of the government. The standards set by safe manufacturers and coordinated with the National Fire Protection Association (NFPA) are also recognized as being adequate to protect other types of materials and information.

Classes of Containers

Class I An approved insulated container that will resist twenty man-minutes against forced entry, twenty man-minutes against surreptitious entry, twenty man-hours against manipulation of the lock, and twenty man-hours against radiological attack. Must have nonmetallic wheels.

Class II An approved insulated security container that will resist five man-minutes against forced entry, twenty man-minutes against surreptitious entry, twenty man-hours against manipulation of the lock, and twenty man-hours against radiological attack. This type normally has metal wheels.

Table 9.4 Sample Identification Label
1. Cabinet Model
2. Serial Number
3. Year of Manufacture
4. Contract Number

Class III An approved *noninsulated* security container that will resist *zero* man-minutes against forced entry, twenty man-minutes against manipulation of the lock, and twenty man-hours against radiological attack. (Class III containers have been removed from the Federal Supply Schedule and replaced by Class V and Class VI containers). Class III containers already in use or on-hand may be used in lieu of Class V and VI containers until replacement is needed.

Class IV An approved noninsulated security container that will resist five man-minutes against forced entry, thirty man-minutes against surreptitious entry, two man-hours against manipulation of the lock, and twenty man-hours against radiological attack. (Class IV containers have been deleted from the supply schedule and replaced by Class V and VI containers).

Class V An approved noninsulated security container that will resist ten man-minutes against forced entry, thirty man-minutes against surreptitious entry, twenty man-hours against manipulation of the lock, and twenty man-hours against radiological attack.

Class VI An approved noninsulated security container that will afford *zero* man minutes against forced entry, thirty man-minutes against surreptitious entry, twenty man-hours against manipulation of the lock, and twenty man-hours against radiological attack.

Vault Doors The above protection applies only to the door assembly and not the entire vault.

Note that only Classes V and VI are in use today. The previous classes of containers should not be used. The exception to this rule within government is that any of the previously approved GSA containers that have been in *continuous use* may continue to be used until they are taken out of service. This means that even if a container is emptied and turned in to the organization supply room for only one day, it cannot be reissued to another office for certain types of storage.

Combination Lock Groups

Group 1 Lock A U.L. classification for combination locks having a possibility of at least 1,000,000 combinations and will resist manipulation for at least twenty man-hours. A Group l-R combination lock will also resist radiological attack for at least twenty man-hours. The group marking on combination locks is found on the rear faceplate or cover of the lock.

Group 2 Lock A U.L. classification for a combination lock that has a possibility of at least 1,000,000 combinations but is susceptible to manipulation by skilled personnel. (The Group 2 lock, with metallic wheels, will not be used for the storage of government classified information.)

Combination Padlocks There are only a few models that fall in the category of changeable combination padlocks that are currently in use and were manufactured and procured under federal specifications by GSA. All four models are manufactured by Sargent and Greenleaf. It is possible that with improving state-of-the-art technology in locking devices, others will soon become available.

Model 8088 This model has been replaced by Model 8077A, which is described below. There are several models of the 8088, but none are approved for use on security containers.

Model 8077 The 8077 is similar in physical appearance to the 8088 but has a chromed lock case. Most of the dial is concealed, except the portion near the number being dialed. The combination change hole is in the rear of the lock, concealed by a sliding back-plate. This lock is serially numbered both on the shackle and on the rear of the lock to prevent lock substitution. This particular model is no longer approved for use on security containers, but other authorized 8077 models still maintain the serial numbering system.

Model 8077A, AB, or AC This lock is similar to the 8077 model except for the modification of the sliding back cover. The back cover is manufactured in heavy steel plate, which provides additional security not provided by the 8077 model. This lock is approved for use on security containers.

Model 8077AD In mid-1995, the S&G 8077AD was approved. This most recent combination padlock has dialing tolerances tighter than any previous models. Because of this, the dial must be viewed straight on rather than from the side or even from a slight angle to ensure accuracy. This becomes more critical when changing the lock's combination.

Model 8065 This is a special padlock without an external shackle. It looks somewhat like the model 8077 and 8077A without a shackle. This lock is designed to be used in situations where the lock can be placed flat against the locked surface over a suitable locking staple.

Moss-Hamilton X-07 This is a somewhat recent combination lock (1993) that operates on a principle not used in any other locking device currently on the market. It has an interior computer chip and operates on battery power that is generated when the dial is first rotated. It is a unique combination lock and, as of 1993, is the only authorized lock for new containers or lock replacement. The cost is several times more than the S&G model. As of this writing, the X-07 is the only lock that meets government combination lock standards.

In 1994, the Sargent & Greenleaf Company developed the Model 6120 motorized, electronic combination lock. This is a tamper-proof, electronic-controlled lock that has proven to be 100 percent resistant to vibration attacks, unlike some solenoid or mechanical locks. It provides the security normally reserved for a U.L. Group 1 lock.

As of 1995, S&G has developed a competitor to the Moss-Hamilton X-07. Similar to the X-07, it is unique, full of computer electronics, and is now in the testing phase. It is expected to be priced somewhat lower than the X-07. Using the "hands-on" technique to compare it to the X-07, the S&G competitor was determined to be easier to use, easier to set or reset, and less complicated to figure out.

Storage Vaults and Rooms

Storage vaults for use in the storage of classified information are used in all types of facilities. Acceptable vault construction standards are shown in Table 9.5.

When considering purchasing and installing—or converting a room to—a storage vault, consider facility insurance requirements. Insurance companies will set certain construction and protection standards commensurate with those of banking institutions for the protection of certain types of valuable documents or other information, to say nothing of monetary instruments.

Table 9.5 Minimum Construction Standards for Class A, B, and C Vaults

Vault Classification	Approved Storage Level	Floor	Thickness Walls	Ceiling
A	Top Secret	8" RC[1]	8" RC	8" RC
B	Secret	4" RC	8"[2]	4" RC
C	Confidential	4" C[1]	8"[3]	4" RC

RC = Reinforced concrete
C = Concrete without reinforcement and C vaults

NOTES:
1. All concrete used in vault construction will be monolithic cast in place, conforming to a minimum compressive strength of 3,000 psi after 28 days of aging. Reinforcing will be by a minimum ⅝-inch diameter steel reinforcing bars (rebar) laid to a maximum of 6 inches on centers, creating a cross-hatched steel curtain, to be sandwiched at half thickness of the concrete, parallel to the longest surface. Rebar will be anchored or embedded in all contiguous walls/surfaces.
2. Class B vault walls will be constructed of masonry at least 8 inches thick, such as brick or concrete block employing adequate bond. Hollow masonry, only of the vertical cell (load-bearing) type, can also be used, but if used, each cell will have from ceiling to floor ½-inch diameter or larger rebar inserted, and then be filled with pea gravel and Portland cement grout. Rebar will be anchored in both floor and ceiling to a depth of at least 4 inches. In seismic areas, 6 inches or thicker RC will be required.
3. Class C vaults will be constructed of thick-shell concrete block or vertical cell clad tile and be not less than 8 inches thick. In areas of somewhat frequent seismic activity, 6 inches or thicker RC should be used.

Storage rooms or areas converted to storage must meet several acceptable minimum standards. Such rooms are usually within security-controlled access areas, thus they do not follow the standard storage vault criteria. Such rooms, with some exceptions, are not used for open storage of classified material.

Rooms or open areas within a facility controlled area that are converted to storage rooms for classified material will have the following standards applied:

1. True floor to true ceiling walls, appropriately covered so that an attempted penetration can be easily discovered.
2. A solid wood or metal-clad wood door.

3. A three-position dial-type approved combination lock mounted on the door.
4. An interior intrusion detection alarm system sensor(s).

Areas that have been authorized for the *temporary* open storage of classified material will also meet the above standards. All such areas should be visually checked at least once every ninety days by a security office representative. A continuing need for open storage will be resubmitted and rejustified every twelve months for approval or disapproval by facility security office personnel.

Within the subject area of security containers, vaults, and classified storage rooms, physical security personnel must also evaluate the responsibilities and precautions with regard to the custodian/supervisor. For the protection of information and other material, it is important to consider the custodial responsibilities, care during and after normal working hours, and the designation of security container combinations.

Custodian Responsibilities

Custodians are responsible for providing protection and accountability of information at all times and for ensuring that such information is locked in appropriate security containers whenever it is not in use or under the direct supervision of an authorized individual. The custodian must follow prescribed regulatory guidance, including facility policy and procedures, to ensure that unauthorized persons do not gain access to such information.

In government areas, only the activity or facility head (or the assigned designee) may authorize the removal of classified information from designated working areas. Check appropriate regulations to ensure maximum protection of the material; the removal is permitted only if operational requirements exist for its removal. Personal convenience on the part of an employee is not justification for approval of a request to remove information from the facility. Each authorization request for removal must be handled individually. Finally, authorization will only be granted when the materials can be protected under adequate security safeguards.

At government installations, the responsibility for protecting sensitive, unclassified or classified material is the responsibility of each and every individual. Security regulations and policy do not guarantee protection and cannot be written to cover all situations. Thus, a logical interpretation of the regulations must be applied using common sense and security principles. Recently, government regulations have changed such that an agency head must authorize the removal of classified information; this is under much more stringent rules than before.

Care During Working Hours All personnel shall take appropriate precautions to prevent unauthorized access to classified or unclassified sensitive material. Classified material removed from a security container shall be kept under constant surveillance and should be face down or covered when it is not in use. Appropriately designed classified document cover sheets shall be used. These are attached to the face of a classified document whenever it is removed from a security container.

Preliminary drafts, carbon sheets, plates, stencils, stenographic notes, worksheets, typewriter ribbons, and other items containing classified or other sensitive information shall either be destroyed immediately after having served their purpose or shall be handled and protected equivalently to the information they contain.

Destruction of typewriter ribbons, printer ribbons, personal computer diskettes, and the like, from which classified or sensitive information can be obtained, shall be protected in the same manner as the information they contain, or else they shall be destroyed by burning.

Although we are in the computer age, many people keep old typewriters. Keep in mind that after the upper and lower sections of a fabric ribbon have been cycled through and overprinted five times, they may be treated as unclassified, regardless of their classified use thereafter. Carbon and plastic ribbons and carbon paper that has been used in the production of material should be destroyed, but further considerations of future technology must be considered also.

Currently, any ribbon that uses a technology that enables a ribbon to be struck several times in the same area before it moves to the next position may be treated as unclassified/nonsensitive. However, in the not-too-distant future, a technology will emerge that allows such ribbons to be read or deciphered, thus rendering this policy unworkable. Whenever possible, therefore, typewriter ribbons that fall into this category should be considered classified and handled as such. Destruction of the ribbon by burning, shredding, or some other means will ensure that information cannot be reconstituted.

End-of-Day Security Checks A system of security checks must be established for each facility at the end of the work day. This ensures that the physical protection of all classified or sensitive classified material is taken into consideration and appropriate protective measures are applied:

- All classified material must be stored in approved security containers.
- Burn bags are properly stored or destroyed.
- Wastebaskets do not contain classified material.
- Desks, tables, and/or other work areas are free of classified and other extraneous, but related, materials. (A "clean desk" policy is always best.)
- All security containers have been locked by one individual and checked by *another* individual.
- The entire office area for each portion of a facility has been given a complete visual examination to ensure that no classified or sensitive unclassified material has been left out.

Security Container Combinations and Designations

Security containers should be individually numbered for control purposes. There should be nothing on the outside of the container that indicates what type(s) or level(s) of information is stored in the container.

Security container combinations must be changed only by the individuals who have the responsibility and appropriate level security clearance. Combinations shall be changed:

1. When placed in use.
2. When an individual knowing the combination no longer requires access.
3. When the combination or container has been subject to possible compromise.
4. At least annually, unless other procedures or regulatory guidance dictate otherwise.
5. When the container is taken out of service. (At such times, built-in combination locks shall be reset to a standard combination of

50-25-50, while combination padlocks shall be reset to a standard combination of 10-20-30.)

Security container combinations, including vaults and storage rooms, shall be assigned a security classification level equal to the highest level of material authorized to be stored in them. A record must be maintained of the combination for each vault, secure room, or container. The record will have the location of the container; the name(s), home address(es), and telephone number(s) of the individual(s) authorized access; and the date the container combination was last changed. The dissemination of the container combination will be restricted to those individuals with a valid need to know. A master combination storage area within a security container that belongs to the facility security office is a true necessity—it ensures proper control of the combinations that allow access to all other security containers within the facility.

Custodians, not to mention the facility security personnel and personnel of affected offices, must realize that electrically actuated locks (cypher and magnetic strip card locks) do not afford any reasonable measure of true security protection. These types devices should never be used as a substitute for a proper locking device or for securing company-sensitive or government classified information. (Cypher locks and the like are only *a temporary delay device*, **never** a true lock.) This way of thinking must apply even to short periods of time when the office may be vacant and individuals are "just down the hall a moment."

10
Intrusion Detection Systems

The use of an intrusion detection system (IDS) for the protection of a facility, like all other systems and devices discussed, is only one facet of the overall physical security program. The use of IDS is based upon the premise that such devices can detect and/or announce the proximity of a person or an attempted intrusion into an area, past a certain point, or within an area. An IDS alerts the security guard force personnel to an action that endangers or may endanger the security of the facility, a portion thereof, specific areas within, and/or certain equipment and protected information within the facility area.

While the primary means of protection is by means of personal observation, the use of IDS devices should never be ruled out. Personal observation is generally limited to the actual times that security force or other responsible personnel are monitoring specific areas of the facility. IDS devices are a valuable supplemental security aid when continuous surveillance is required, either during working hours, nonworking hours, or both.

Intrusion detection systems are actually protective alarm systems, and are used to accomplish one or more of the following:

- To permit a more economical and efficient use of security guard manpower by substituting fixed security post devices that will provide continuous monitoring capability in place of a larger number of security force personnel. This allows security personnel to perform other security functions while still maintaining a direct interest in the security of each fixed location.
- To provide additional controls at sensitive or vital areas as an additional "insurance" feature against human, electrical, or mechanical failure.
- To substitute for other necessary elements of a physical security program that cannot be used because of the facility or building design or layout, safety regulations, activity operating requirements, cost and budgeting factors, or because of other locally controlled factors.

The decision to use IDS in such applications for added security should be based on:

- The vulnerability and sensitivity of the protected facility.
- The degree of security protection required at any given point or points.
- Other security support currently in use (including support provided by a higher authority of responsibility, such as an installation commander, local police, or private security patrols).
- The availability of manpower for the facility security guard force.
- The cost effectiveness of such a system.

The advantage of an IDS system is its ability to measure reliability of a given protected point. This aspect of any IDS system should never be overlooked, because it is the prime reason to have the system in operation.

There are a wide range of complexities between various types of IDS systems currently available, with newer systems and individual component devices being developed or upgraded constantly. Each can be tested and individually evaluated to determine what degree of security can be expected, and in what manner each type of device should be employed for maximum security protection. The choice of detection device must be based upon what must be most readily detected in a given situation. In many cases, more than one type of detection device should or could be employed for maximum security protection.

From an operational standpoint, the application of an IDS consists of equipment and devices that fall within two areas:

1. Detection. This is a basic subsystem for any IDS, and encompasses the use of sensor devices, a sensor data transmission line, and a display segment from which the device can be continually monitored. Sensors that employ various techniques to detect the presence or movement of personnel and/or vehicles are of wide and varied types. The data transmission line for the device *must* be a hardwired, protected line to ensure security.
2. Surveillance. Surveillance uses various techniques to show an area or specific location within the facility on the monitor. Surveillance may cover a wide area (room protection) or a very small area (point protection).

Because the use of alarms has become a necessary part of a truly integrated physical security system, their application is required in instances where protection is of critical importance to an area or facility. Local situations and conditions that pertain to the facility or are mandated under contract (including insurance requirements) may determine their need. Those conditions and situations that are unique to the facility will, of course, affect the ultimate decision of the type of IDS devices that will be used.

In general, the following should form the basis for determination of alarm use:

- The critical importance and vulnerability for any restricted or controlled access area or facility. This can include the concentration of vital components, materials, and/or data.
- Critical or facility activity-sensitive practices or procedures and controls on certain types of processes in effect.
- Areas in which access is maintained from inside the facility or area and the use of security forces is just outside this area.

Security force personnel may or may not be authorized routine access to the area except under extreme or emergency conditions.
- The remoteness of the area or facility is such that a security guard will not be constantly on site; thus, a remote site monitoring capability is more advantageous (and cost effective).

FACTORS AFFECTING THE OPERATION AND USEFULNESS OF AN IDS

Alarm systems, because of their cost, are justified only when their use results in a commensurate reduction in other costs or when the need absolutely dictates a level of security protection that cannot be obtained otherwise. To afford a required degree of protection that is acceptable to management, and also the degree that will use appropriate devices designed for that specific purpose, the following factors must be considered:

- The known capabilities *and limitations* of the system.
- The design of the facility and internal controlled and monitored locations that the system will be used to protect.
- The effects that weather, air conditioning, heating, adjacent traffic noise, vibrations, and the like will have on the system components and their reliability.
- The effect of the system on day-to-day operations of the facility.
- Whether or not the system will be secure itself against undetected intrusion from any reasonable approach to the facility or a specific area that would allow for tampering with the system.
- The system must be designed so that the time interval between the sensor's detection of an unauthorized activity by an individual and the time allowed for response is such that the objective of an unauthorized activity is not compromised.
- All systems, materials, and equipment meet standard specifications and requirements. In this regard, consider the Underwriter's Laboratories as the standard to meet or exceed for the system and all individual sensor devices used. All such equipment should have the U.L. seal affixed.

Other system characteristics that must be known prior to the selection of a system include the following (if system is already in use, then all parts of the system must be reliable and able to meet these criteria):

- If it is a local site-monitored system, the protective circuits and devices must be connected to a visual and/or audible monitor element, which is located near or in the immediate vicinity of the protected facility, and is monitored on a twenty-four-hour-a-day basis by trained security force personnel.
- If the system is a central station type (where one or several facilities are provided protection from the same monitoring point), the circuits and devices must, when activated, automatically signal the central station, which has a response force of trained security personnel. The central station monitoring facility must be manned twenty-four hours a day by personnel who are trained and know the strengths and limitations of the devices maintained at the various facilities.

Knowing the type of monitoring station capability and its location is only a small part of system determination (for selection of a system or one currently in use) with regard to site-specified security protection. Such items of information should be common knowledge among the security inspection team members.

Regarding the sensor devices in general, as well as the overall system, desired—if not required—features include:

- Capability to detect any attack or attempted intrusion upon doors, windows, walls, floors, or ceilings of storage facilities or individual points of a perimeter to an area or facility. For spot protection, the capability to detect any attempt to enter into or move the device or object being provided such protection.
- A high degree of salvage value (it can be removed and installed elsewhere).
- The system and individual devices should operate on 110 to 220 volts, 50 to 60 cycles, AC power, and have a self-contained automatic switch-over capability with an emergency backup power source that allows for continuous emergency operation lasting six hours or more.

- The alarm annunciation panel board located at the main security guard desk to facilitate prompt response to an alarm condition.
- The system must have a built-in safeguard to annunciate an alarm at the panel board when the mode of operation is changed at the sensor control box.
- A reasonable flexibility to protect both large and small areas with small cost differentials.
- Have both an audible and visual alarm annunciation capability at the main alarm console panel.
- The false alarm rate for the IDS does not exceed more than one per thirty days for each alarmed sensor in operation.

IDS system performance, in addition to the above concerns that may affect the system, must include those features that relate to the system itself or factors pertaining to the control, maintenance, and/or proper monitoring of the system:

- There must be an effective, regularly scheduled maintenance and system check in operation; the maintenance contract must call for a timely emergency response to system problems. Regularly scheduled maintenance and testing of the system should occur at least once every sixty days (thirty days being preferred) and emergency response under ninety minutes to any given emergency call.
- All personnel monitoring the alarm system must be fully trained to understand the system, including operation of individual sensor unit control panels and basic fault isolation.
- The IDS system must be protected against bypassing or spoofing. This can be accomplished by having individual sensor control panels for various areas maintained within the area being protected. Junction boxes outside protected areas, as well as the sensor control panels, must have a tamper switch inside panels and panels secured (outside area panels must have a padlock device to secure them). Transmission lines must have wireline protection against tampering. The line supervision must be able to detect bypassing, spoofing, interception, or cutting of lines. All line supervision should be commensurate

with the value, sensitivity, and mission criticality of the protected resources.

Note that simple line voltage testing, current variance, and line resistance systems offer the *least* amount of protection.

- When possible, data transmission systems should employ random signal generators because this provides a more sophisticated line supervision capability.
- Facility and/or individual areas being monitored should be prominently marked using signs indicating that the facility is protected by an IDS.
- If the wireline runs outside a given facility building, protect the transmission lines against dislocation, breaking, cutting, or other natural or manmade hazards. Transmission lines should be protected in metal conduit and buried at least twenty-four inches below ground surface or suspended at least fifteen feet above ground. Wirelines at or near perimeter barriers or fences or in uncontrolled areas should be underground, never above ground.

TYPES OF EQUIPMENT AND PROTECTION

Each year newer types of alarm systems and individual sensor devices enter the marketplace. Some of these can be classified only as gadgets or novelties with a very restricted use, but others will reach the maximum level of complexity. Over the years, though, a number of types have successfully withstood the test of time and now form the backbone of the alarm industry. Only the salient features of each type will be covered in brief detail to provide a better understanding of their application within the inspection program.

In addressing the various types of system sensor devices in use, bear in mind that they are used for three types of protection:

1. Point or Spot—protection of security storage cabinets, files, or very small areas or rooms within a larger room or building.
2. Area or Space—protection of a portion or the total interior of a room or building.

3. Perimeter—protection of exterior doors, windows, and/or other openings or the protection of fences, gates, and other outdoor accesses.

Any or all of these types of protection can be used with the security systems previously discussed.

Electromechanical Sensors

IDS sensors are either electromechanical or electronic. An electromechanical sensor consists of a continuous electrical circuit that is balanced so a change or break in the circuit will set off the annunciator (alarm) at the monitoring panel. Currently, the following sensors fall into this category.

Window Foil or Tape This is a metallic tape affixed to the interior side of windows and glass doors. When the glass is broken, the foil breaks, an open circuit results, and an alarm is sounded. A hairline crack or scratch will also activate the system, causing an alarm to be sounded.

Wire Lacing and Screening This electromagnetic device uses wires laced across door panels, floors, walls, and/or ceilings. A forced entry into the protected area breaks the wire, causing the alarm to sound.

Taut Wire A taut wire device is used to detect intrusion into a protected area. A wire under tension is strung across an opening or space such as air ducts or utility tunnels. Any change in the tension of the wire causes an alarm to sound.

Contact Switches These are magnetic or mechanical intrusion devices and are frequently used to protect doors, windows, skylights, and other accessible openings. Switches may be surface mounted or recessed.

- Magnetic Intrusion Switch—This switch consists of two separate parts, one a magnet and the other a switch assembly. When the magnet is properly oriented and mounted adjacent to the switch assembly, the switch is in an activated mode. When the magnet is removed, or moved farther away from the switch, the switch is deactivated, causing an alarm to sound. Normally, the magnet is mounted on the movable portion of a door, window, or other item to be protected.

- Mechanical Intrusion Switch—This switch is also activated by opening a door, window, skylight, and the like. The plunger-type switch is normally recessed. The lever-type switch is less expensive but is more easily detected. Mechanical switches exposed to weather may stick or freeze and become temporarily inoperable, allowing access without an alarm being activated.

Vibration Detector   This device is a self-contained switch in a small housing. One contact of the switch is a small pendulum-type weight, which is usually held under a slight spring tension touching the contact and maintaining a continuous electrical circuit. An attack on the surface results in vibrations, which, in turn, cause the pendulum to swing away from its normal position. This opens the circuit and activates the alarm. The vibration detector can be used as a small measure of protection for attack through walls, ceilings, and large glass surfaces.

Electronic Sensors

The following comprise the commonly used electronic sensors.

Photoelectric Detector   A photoelectric detector is designed to transmit a beam of light from a mounted light source to a light-sensitive receiving unit, which will react to a decrease or elimination of light. This reaction results in the initiation of an alarm signal. The components are usually arranged so that the beam of light crosses the approach to the area being protected. Photoelectric detection devices can be used indoors or outdoors. To project a light path other than a single, straight narrow beam, mirrors may be installed. By using such mirrors, the beam may be zig-zagged back and forth to make it much more difficult to avoid. The effective range of the device is decreased with mirrors because of the light loss through reflection. As a protective alarm device, the photoelectric detector can be considered little more than supplementary.

Proximity Detectors The proximity detector operates by surrounding an object with a balanced electrical field; entry of an electrically conductive body creates an imbalance in the system and triggers the alarm signal. For proximity detectors, the device or object being protected should be mounted on a rubber mat or other nonconductive covering. The covering should be completely under the device and extend at least twelve to eighteen inches away on all sides at floor level.

Motion Detection Devices The basis for the operation of these devices is the Doppler effect. When the source of a sound or electromagnetic signal, or a reflector of such a signal, moves away or toward the receiver, the frequency or pitch of that signal being received will be higher or lower. The motion of an individual within such a protected area causes a change in the wave energy frequency of the waves received. The receiver picks up the source frequency but also detects a slightly different frequency at a much lower strength—possibly indicating a very slight body movement by an intruder.

- Sonic. Sonic detection systems operate in the audible range, 1,500 to 2,000 hertz and higher. The constant tone is annoying because it is within the human audio range and at a high-decibel output. This system uses transducers (transmitters and receivers) to saturate the enclosed area with sound waves. The transmitting and receiving transducers are permanent magnet (PM) speakers mounted in the same room, usually on opposite walls. The receiver listens to the tone being transmitted and compares the reflected signal. Whenever the pattern of the tone varies because of a disturbance within the protected area, the receiver will detect this change in frequency and activate an alarm.
- Ultrasonics. Ultrasonic detection utilizes high-frequency sound waves at greater than 20,000 hertz. Otherwise, it is much like a sonic system. Because the frequencies are at the upper level of the audible range, only a few persons (generally children) can hear them.
- Microwave. Microwave operates in a similar manner to the above systems, but they are transmitted at a frequency between 400 and 10,000 hertz. The signals can be controlled with respect to the size of the area to be protected through the selection of the antenna. One of several antennas can be used

to provide the required protection without interfering with sonic or ultrasonic units.

Acoustical (Audio) Systems These systems utilize a type of microphone to detect sounds that exceed the ambient noise level of the area under protection. Obviously, such devices cannot be employed in areas where they can receive noise from manmade sources, such as construction, aircraft, excessive noise, walking, talking, or aircraft. Such noises will set off a nuisance (false) alarm. Some of the more sensitive system devices can even be triggered by sounds of rain or thunder. While some acoustic systems rely upon air to transmit the sound to the detector, others will not respond to ordinary noises in the air but only to those transmitted through a structure such as a wall.

- Acoustical (audio). An audio detection system listens for intrusion sounds by using microphones installed within the protected area. Upon detection, an alarm occurs. This type of system may be equipped with cancellation and discrimination units that electronically evaluate the significance of the sound disturbance, thus eliminating reaction to nuisance alarms caused by airplanes, thunder, and so forth.
- Vibration (seismic). This type of system utilizes the same principle as the audio detection system except that highly sensitive and specialized microphones are attached directly to objects such as safes, filing cabinets, windows, walls, and ceilings. Vibration of these objects initiates alarms. Cancellation and discrimination units are necessary to counteract nuisance alarms.

Proximity Systems Proximity detection devices operate by surrounding an object with an electrical field in such balance that the entry of an electrically conductive body into the field creates an imbalance, activating the alarm.

There are several methods for establishing the magnetic field; methods will differ to some extent among system manufacturers. A proximity system may also be employed to protect an area by erecting what is commonly known as a magnetic fence, an integral part of the system. Other variations provide surveillance of doors and windows.

The proximity system is designed to be supplemental and cannot be used effectively as a primary system. This is because the system is susceptible to nuisance alarms caused by electric supply fluctuations and by metallic materials that are placed too near it. Animals and birds can also trigger a system into alarm if it is too sensitive.

Summary Detection devices are usually designed to detect a single phenomenon. The type of detection device chosen is based upon what phenomenon will be the most readily detectable in a given situation. It may be desirable to employ more than one type of detection device to protect against all possible methods of entry. Usually, similar equipment is manufactured by several different companies; it will operate on the same basic principles but may differ in refinements, and these differences may alter the degree of security.

The most common detectors seen during inspections are the electromechanical and motion detection devices. The next most common are photoelectric sensor devices.

Table 10.1 compares more commonly available sensors with regard to their type, limitations, and resistance to defeat. Table 10.2 compares IDS sensors by type of sensor.

Table 10.1 Comparison of Currently Available Sensor Devices

Sensor Type	Limitations	Resistance to Defeat
Photometric	Extraneous light must be excluded from area; limited to interior rooms.	High
Motion Ultrasonic	Air motion may cause false alarms.	Moderate to High
Motion Microwave	Energy can penetrate walls, etc., causing nuisance alarms.	High
Acoustical-Seismic, Sound	Extraneous noises will generate a nuisance alarm.	High
Acoustical-Seismic, Vibration	Localizing the source of nuisance alarms can be difficult.	High
Proximity, Capacitive	Susceptible to nuisance alarms; requires backup.	High

Table 10.2 Comparison of the Advantages and Disadvantages of IDS Sensors by Type of Sensor

System Type	Advantages	Disadvantages
Ultrasonic	Provides effective protection against intruders concealed within premises.	May require reduced sensitivity to overcome possible disturbance factors in enclosures (such as telephones, machines, clocks, etc.)
	High salvage value	Can be set off by high external noises.
Microwave	Protective field not visible; therefore, difficult to detect presence of, or compromise to, the system.	
	Good coverage provided if antennas are properly placed.	Coverage not easily confined to desired area. Penetrates wooden partitions and windows, and may be accidentally activated by persons or vehicles outside the protected area.
Capacitance	Not affected by air currents, noise or sound.	Fluorescent lighting will activate sensor.
	Extremely flexible; used to protect safes, cabinets, doors, windows, etc. Any unguarded metallic object within maximum tuning range can be protected.	Can only be applied to ungrounded equipment.
	Simple to install and operate.	Accidental alarms occur if protected area or object is carelessly approached, such as by cleaners at night.
	Provides invisible protective field, making it difficult for intruder to determine when the system has been activated into an alarm condition.	Housekeeping of protected area on object must be carefully watched.
	Easily dismantled and reinstalled.	
	Compact equipment.	
	High grade of protection.	

Typical Sensor Applications

Exterior Doors Sensors on exterior doors guard against unauthorized entry. One or more balanced magnetic switches may be used. Also, the surface of a wall or interior side of the door can be covered with a grid wire sensor (commonly referred to as lace and cover) using the principle of breaking an electrical circuit.

Solid Walls, Floors, and Ceilings These areas are covered with grid wire sensor, or else the room is equipped with a passive ultrasonic sensor. Sound detection and vibration systems may also be used to detect penetration through such areas.

Open Walls and Ceilings Wire cage walls and ceilings present distinct problems. Certain modifications are necessary to protect this type of construction. The walls and ceiling may be enclosed with building materials on the outside of the protected cage, thus, permitting the use of ultrasonic or grid wire sensors within the cage area.

Windows Whenever possible, windows should be eliminated. Where necessary, consider use of interior metal shutters. If the character of the room does not allow for the use of a passive ultrasonic sensor, then a vibration, capacitance proximity, or other sensor can be used.

Ventilation Openings Cover openings with steel bars, mesh, or louvered barriers. Intrusion through ventilation openings can be detected via a passive sensor or vibration sensor.

Construction Openings These are unsecured openings from incomplete construction and should be covered with a grid wire installed on plywood. Where the opening must stay open, a capacitance proximity sensor should be used inside the opening.

Applications for Vaults

Security vaults are used for the storage of highly sensitive, controlled, and classified storage-type applications. In large part, they are used for mail and distribution activities, but they may also have different applications, depending upon their location and purpose.

Vaults *need* to be protected by an intrusion detection system. Where the installation or facility has an existing IDS serving other sensitive or classified activities, that system can be extended to monitor the operation of the intrusion detection devices installed on the vault. In such cases, a supplementary, local audible alarm on the exterior of the vault is optional. At places where there is no existing system, but guard personnel are constantly in the vicinity of the protected space and are capable of responding to an alarm within three minutes or less, a "local" IDS can be installed in the protected area. Where guard personnel are not as available, operation of intrusion detection equipment installed on the protected space shall be monitored by electrically supervised lines that extend to the nearest, continuously manned police or security force operations center or to a civilian police station capable of responding to an alarm. (In some localities this arrangement will not be feasible because the civilian police are not willing to monitor proprietary IDS.) Manually operated silencing switches on exterior audible alarms shall not be installed outside the vault.

Detection devices installed to protect vaults can be of the type indicated below (all are mandatory):

- A balanced magnetic switch on the inside of the door
- An interior volumetric detector
- A capacitance detector on any interior safe
- Ultrasonic motion detector, vibration sensors on interior of vault. Note that there can be some minor variations, depending on the availability and uniqueness of the facility location.

The arrangement of circuits for vaults can be in two configurations. Where detection devices are monitored at a guard station, not in the immediate vicinity of the protected space, the detectors shall be arranged in two circuits—one to monitor perimeter detectors and the other to monitor interior detectors. Where a "local" system only is installed, all detectors will be connected into one circuit.

MILITARY APPLICATIONS OF THE JOINT SERVICE INTERIOR INTRUSION DETECTION SYSTEM

The Joint Service Interior Intrusion Detection System (JSIIDS) is the main system available for military applications. It has numerous applications, just as it has an extensive amount of various equipment. Other similar systems are available and are being brought into use. Any IDS that has been installed previously, or that will soon be installed, that falls within the JSIIDS is on an approved products listing by the U.S. government. Such systems have been approved for installation and provide—or can be modified to provide—a comparable level of protection equal to or better than the JSIIDS.

In all cases, the "on/off" switch that activates the alarm system must be located within the alarmed area. The facility security office will evaluate existing IDS for adequacy at least once a year, while new installations of IDS applicable to the JSIIDS concept should be evaluated at the three-, six-, and nine-month levels after initial installation.

The IDS shall include a central control station and shall provide for a response at the scene not later than three minutes after an activated alarm. For facilities that do not maintain a round-the-clock security guard force, but rely upon an installation or local police response, response time should not be more than five minutes from the time an alarm is activated.

In cases where the IDS is used for civilian communities and private industry applications, and when an on-site security guard force is not available, arrangements should be made to connect alarms to police headquarters (if authorized) or, preferably, to private security alarm companies from which an immediate response can be made or directed. Response time should not be more than ten minutes after initial activation of an alarm. A daily log will be maintained of all alarms received and should include the nature of the alarm, whether it indicated an intrusion system failure or nuisance alarm, and, as a minimum, the date and time the alarm was received, the location, and action taken in response to the alarm.

Transmission lines for all alarm circuits shall be supervised and dedicated to preclude undetected tampering. A protected, independent power source must be provided as a backup for the system.

11
Closed-Circuit Television

Closed-circuit television (CCTV), although not an intrusion detection system, is an extremely valuable adjunct to the overall physical security program and the day-to-day operations of a facility. It is frequently used to complement the facility's security program, by supplementing the facility security guard force.

CCTV cameras support the security program by being placed at critical locations throughout the facility to provide direct visual monitoring from a given vantage point. CCTV may be used on gates, along fence lines where other observation may be difficult, interior hallways, doors to sensitive areas, high-value storage rooms, remote or out-of-the-way areas and buildings, and Automated Information Systems (AIS) sites. When such areas are not continually manned, CCTV offers an alternative to higher personnel costs while still providing an acceptable measure of security, because it can be remotely monitored and controlled.

CCTV, in its simplest form, consists of an unmanned, mounted television camera, a wireline cable, a video monitor, and in some instances, a video recorder. Accessory equipment may include an automatic switching capability to switch viewing on the video monitor for cameras located at various facility locations, a freeze frame capability, and others.

The normal use of CCTV at entry points to a facility should also include the use of a two-way communications system between the monitor site panel and the gate, and an electrically operated gate when required. In the majority of instances, however, certain gates are locked when not in use and thus no CCTV is used to monitor the gate. These gates, and those that are open continuously (or open on a daily basis) should have the CCTV located at an interior point of higher ground where sweeps of the gate and perimeter line can be accomplished continually by a remotely activated, motor-driven camera.

Depending upon the facility, there may be a security requirement requiring an electrically operated gate and two-way communication tied into a CCTV. This eliminates a guard at noncritical times from a main entry gate, because the person at the monitor panel can be alerted on a speaker system by the person wishing to enter. The person at the monitor panel can observe that person on the monitor and determine their authority to enter the facility. At that point, the electrically activated gate switch is used to release the gate lock. A variation of this is the Twin-guard® system, which uses a split-screen monitor to allow viewing of the individual (*and* his or her identification badge) and talking to the individual simultaneously.

CCTV can also be used for surveillance of security-controlled restricted-access areas, high-value items in storage sheds or warehouses, fence lines, movement of cargo, parking lots, infrequently used hallways, and other such areas.

All controls for a CCTV system should be enclosed in a metal housing and properly secured to preclude tampering by unauthorized personnel. Ideally, all equipment should be located at the security monitoring post.

Cameras should be left in the "on" position at all times. Delay caused by camera warm-up and adjustment can be eliminated by maintaining a continuous CCTV operation.

Normally, surveillance CCTV is of a low light level and can operate under marginal light conditions. This is a key consideration for the selection of a CCTV system in addition to the supportive artificial lighting that is maintained for and within the facility.

Various lenses and accessories should be considered to maximize the capabilities of the CCTV system and to allow the facility security office to obtain maximum and efficient use of the equipment. These items should be looked at during the physical security inspection.

The camera can be equipped with several different types of lenses. In addition to the standard camera lens, a wide-angle lens will increase the camera's field of view and a telephoto lens will magnify the desired scene, bringing the viewer "closer." Lenses are available in different focal lengths.

Remote control equipment may be considered when the advantages outweigh the costs. Such items as a pan and tilt pedestal for the camera allow the camera to be positioned in the desired horizontal or vertical position from a remote location. The camera turrets and zoom lens can also be controlled so that routine observation covers a wide field of view; the telephoto lens may be used to provide a closer observation of an area or individual. A scanning mechanism will cause a camera to turn at the rate of two to three revolutions per minute to provide all-around observation of a facility area.

While CCTV has many proven applications and its use is highly recommended, there are also possible disadvantages to using such a system to replace a guard, including:

1. The monitoring screen does not provide as faithful a reproduction of the scene as does direct vision. For this reason, small details are not discernible or are vague to the eye.
2. Dividing the guard's attention between several monitoring screens may not provide the continuity of coverage desired.
3. The resulting eye strain and the boredom of watching a monitor may cause a lack of attention on the part of the security guard.
4. The area viewed may contain so many obstructions that even several CCTV cameras could not give proper coverage; a roving guard or guard patrol would be a better choice in this case.
5. The camera is incapable of taking corrective actions in response to an event that is taking place. The time required to move a security guard or guards to the area may be too great.

12
Automated Information Systems Security

Automated Information Systems (AIS), when within a facility, must have an equivalent or greater level of security awareness with regard to the rest of a typical facility. Within the AIS environment, many of the operations are designed, developed, run, monitored, tested, evaluated, and transmitted over lines to various remote terminal sites. The computer facility may well be the only reason the entire facility exists.

Within the AIS environment, the PSI team member will be looking at the various areas previously covered in this manual. Such areas can and do also apply to an AIS site, whether it is the site itself or a site within a larger organizational facility.

In general, most security specialists do not have the requisite background knowledge to deal with computer operations and security. Inspection team members will be able to establish their professional credentials by applying their general security knowledge to the physical security inspection of an AIS site. AIS facilities, remote terminal areas, computer rooms, and the like are merely a part of a facility and should be treated as such.

The AIS area does not possess any special mystique, outside of the fact that individual team personnel may neither totally understand nor care to comprehend the intricate details of programming and computer operations. Overall, however, the ability to perform a physical security inspection of the AIS area is more than adequate. This is because team personnel are not checking on the AIS internal controls, but on the physical security protection features for the entire AIS system.

On the following pages are the areas that relate to the security of a computer facility or site. Appendix 2 provides a well-defined AIS facility/site security checklist. This is very detailed and can be used also by the AIS Systems Security Officer (AIS SSO) to view *all* the portions of security. Provide the AIS SSO with a copy of the checklist.

The inspection team must bear in mind some general areas of knowledge. While these seem to be self-evident, a review should be performed by each team member. The areas include:

1. A balanced AIS security program *must* be based on a solid physical security foundation. The physical security measures implemented at any AIS installation are designed and used to protect and preserve informational, physical, and human assets by reducing exposure to various threats that could produce a disruption or denial of services or the exploitation of same by foreign intelligence and/or criminal elements.
2. The risk reduction process is accomplished by minimizing or preventing harmful effects of any natural, accidental, or other malicious event that is detrimental to the facility and in turn the computer facility. A good physical security environment is required to complement the overall automated system security procedures that are effected through various administrative procedures or those measures that are incorporated into the system hardware or software.
3. The computer facility systems security office should ensure that data processing assets that are under their control are properly protected through the implementation of effective physical security measures. If the AIS system is tied to the Department of Defense, it should conform to the requirements of JCS Pub 22. For another government agency, it should use an equivalent publication. For the private sector, a variety of publications that set voluntary standards are available and should be reviewed. Select the one(s) most appropriate to the situation and environment.

PHYSICAL SECURITY PRINCIPLES FOR AUTOMATED INFORMATION SYSTEMS FACILITIES

AIS facilities should have multiple barriers for protection and access control. Such barriers ensure the physical protection of the facility equipment, information, and personnel.

1. An in-depth application of a series of barriers and procedures provides the basic physical security protection to the installation and central computer facility. Barriers and procedures employed to attain this desired level of security include definitive structural standards for the protected facility or area through barrier control/access, key control, lighting, lock application, lock inventory and accountability procedures, alarms, and area access controls.
2. Positive physical access controls must be established to prevent unauthorized entry into the computer facility and/or other critical areas that directly support or affect the operation of computer equipment or the processing of data.
3. Access to the computer rooms and media storage libraries, in addition to various remote terminal sites and backup support function areas, must be permitted *only to those individuals* who frequently require access to these rooms/areas in the performance of their official duties. "Frequent" should mean at least one-half of the time.
4. The effects of natural disasters such as fire and floods can be prevented, controlled, or minimized to the extent economically feasible by the use of detection equipment, extinguishing systems, and well-conceived and tested emergency plans. Off-site backup facilities are one method to ensure AIS continuity.
5. Buildings or facilities selected or designed to house computer equipment will be of sufficient structural integrity to provide—or be capable of being made to provide—effective physical security at a reasonable cost.

Location of the Central Computer Complex

Site selection is a key factor in the establishment and maintenance of a secure operating environment. Ideally, the physical characteristics of any location selected to house an AIS will support the establishment of an effective physical security system at the facility. Architectural design is an equally important aspect of the site selection and security relationship. Physical provisions for restricting access should be incorporated into the initial facility design.

1. A location within a building should not be easily accessible to the general public.
2. The equipment room should be located in the center of the building to provide the maximum protection inherent in the building itself. The environment immediately above, below, and adjacent to the facility must also be considered.
3. The requirement to provide protection for other critical areas, such as media libraries, data preparation areas, and environmental support equipment, must also be considered in site selection and facility design.

Access Controls

Physical and procedural access controls are required to prevent unauthorized entry into the main computer facility complex area environs and to control the flow of materials into and out of the facility. Positive access controls should be maintained at all times. This is accomplished via the following means, which are used in conjunction with each other to assure the maximum physical protection:

1. *Designation as a Controlled Area.* The computer facilities, alternate facilities, remote site locations, and specialized areas have to be designated as a form of a "controlled access area." Ideally, all rooms within the facility that are associated with the computer operations, such as the programmer offices, supply rooms, and scheduling tape library, will be included within the controlled area. The computer room and tape media library will also be treated as a maximum-security zone within the Controlled Area.
2. *Personnel Access During Operational Hours.* Appropriate security controls must be developed and implemented to ensure that only authorized personnel are permitted to enter the computer room. This can be achieved through the use of one or more of the following, in combination with other access con-

trols that will maximize the physical and personnel security aspects for the facility: physical barriers, such as counters or locked doors equipped with an electrical release or cypher lock; a receptionist; a badge system, which may or may not include a badge exchange process; or a security guard. It is highly desirable and recommended that a buffer or initial access (outer area) entry be created adjacent to the primary entrance to the computer room. Any use of secondary or emergency entrances will be strictly controlled and monitored through the use of appropriate security system equipment and devices.

3. *Security of Computer Rooms During Nonoperational Hours.* All computer rooms will be secured at the end of the work day or at any other time the facility is unoccupied, such as during a fire drill (when all personnel must actually evacuate the facility), actual fire, bomb threat, and the like. Strict accountability will be maintained for keys, combinations, or identification numbers that permit access to the facility. Entry should be for maintenance or emergency reasons only. Individuals should be regularly known to the security force and be required to sign in, stating the reason for nonoperational hours access. Unknown individuals should be escorted at all times. During an actual fire, the fire suppression systems built into the facility AIS area would be actuated. This avoids opening the area to fire department personnel whose water hoses would damage the AIS equipment. Facility computer rooms have overhead crawl spaces, space beneath raised flooring (which may or may not extend beyond the controlled area of the computer room), and heating or air conditioning vents or ducts that need some form of barrier bars or grillwork installed. During nonduty hours, all entrances to the computer facilities will be protected by a minimum of two separate, independent barriers to provide in-depth security. In addition to locked doors for the computer room (or building, if the computers are housed in a dedicated, separate building), any of the following security measures may be used to provide a dual level of protection:
    - All secondary access doors, physically separated from the first, should be securely locked. Deadbolt locking mechanisms (or the equivalent) shall be used on all locks to

protect the unoccupied facilities. Deadbolt locking mechanisms will have, at a minimum, five tumblers, and such locking mechanisms should not be master keyed. Key control will be positively maintained for these mechanisms.
- If a chain-link fence is used, the entrance should be securely locked. The fence should be at least seven feet in height, topped by an outrigger bar with three strands of barbed wire. Existing fences in good repair that fail to meet these criteria, while providing reasonably adequate security with a six-foot fabric, need not be replaced. Upgrading is preferred, but costs may negate this. If an identifiable threat increases, then this measure must be taken.

4. *Additional Protective Measures.* Should the circumstances warrant (namely, sensitivity, location, and risk), a higher degree of protection—as listed below—may be implemented. This can include:
   - Dedicated guard service, which has appropriate training and meets requisite security clearance requirements, if necessary.
   - All access points continuously monitored by a CCTV in a dedicated mode with secure line protection.
   - Intrusion detection systems (IDS), which may supplement the dedicated guard or CCTV system, to monitor/protect areas, rooms, hallways, entrances/exits that cannot be directly observed by a guard or CCTV monitor or are in somewhat remote areas that are not under a continuing visual monitoring or surveillance.

5. *Visitor Controls.* Appropriate safeguards and physical controls will be developed and instituted at each site location to restrict the access of visitors, equipment maintenance personnel, and all other individuals not directly involved with the management or operating of the computer equipment. It is recognized that different procedures and restrictions will be required for various categories of visitors. All visits by any foreign nationals and other individuals whose identity is not known to the AIS manager should be appropriately coordinated. Packages or other potentially dangerous objects carried by visitors shall be inspected before they are allowed to enter the facility.

6. *Customer Engineers (CEs).* Resident CEs are normally provided office work space that is as convenient as possible to the

computer facility. The assigned area, however, must not afford direct, unrestricted access to the computer room area. All CE work in the computer room will be performed in the presence of a knowledgeable operator (or supervisor) who knows the CE's purpose for being in the computer facility.

Security of Media Storage Libraries

Areas used for media storage of magnetic tapes, disk packs, paper tapes, or card decks associated with the operation of the computer system shall be afforded adequate physical security protection. If the library is outside the protected area of the central computer facility, it will be provided, at minimum, a solid door equipped with a lock approved by supporting security personnel and an IDS device that can be continuously monitored from another location (that is, twenty-four-hour-a-day guard desk). Fire and magnetic field protection requirements unique to this setting, will also be provided as necessary. Any media storage library that is either inside or outside of the protected area should use additional security measures—such as door alarms, IDS, high-security doors, and storage containers—to ensure maximum protection consistent with the level and sensitivity of the information stored within the library.

1. Access to a media library must be positively controlled and limited to personnel assigned to the media library and their immediate supervisors. When the computer facility is not operational or when reduced staffing does not permit effective surveillance to be maintained, the doors to the library area(s) or cabinets that contain classified/sensitive files or programs will be secured.
2. If the media storage area is located outside the main computer facility, it must be constructed of four solid walls extending from the true floor to the true ceiling and be equipped with a solid metal-clad or vault-type door if government-classified material is stored within. This door will be locked after normal duty hours and may be equipped with suitable alarm sensors.
3. An adequate fire protection system will be designed and installed.
4. If there are water pipes within the wall or within the interior side of the wall, water alarms will be strategically placed upon

the floor in, near, and/or about the piping to alert personnel to potential dangers from water leakage. If possible, reroute the water pipes outside the walls and remove old pipes.

Security of Remote Terminals

The introduction and use of large-scale, remotely accessed computer systems requires that equipment be protected from theft, abuse, damage, or unauthorized use.

1. *Terminal Locations.* Areas considered as locations of remote terminal devices should be capable of being secured both during and after normal working hours and should be arranged so that terminal use can be effectively monitored.
2. *Access Controls.* Safeguards have to be implemented to allow only authorized individuals to utilize remote terminal equipment. Minimally, this consists of an official list of authorized terminal users, and of physical safeguards and administrative procedures to assure the restriction of terminal use to authorized personnel. Further, physical access for the terminal must include a locked door with the option of a three-position spy-proof combination lock set in a solid door, a door buzzer to alert personnel already in the room/area that another individual seeks entry, and a cypher lock. If the terminal area is a distance away from the controlled-access area, where continually secure monitoring is not available, then a CCTV camera should be installed with a secure line to the monitoring set to observe the door. An interior IDS system will be maintained with a secure line to the monitoring set at the guard post or other monitoring point. The IDS system needs the capability of day-and-night positions. After working hours or during periods when effective monitoring of active remote access equipment cannot be maintained, these areas are to be kept in a secured mode. Remote terminals should be physically disconnected from the system when not in use. Keyboard control locks and/or other disabling features may also be used, at local discretion.
3. *Accoustically Coupled Remote Terminal Devices.* Because of the increased vulnerability inherent in such terminals, their use in

accessing highly sensitive systems should be avoided whenever possible. When these devices must be used, the following countermeasures need to be implemented:
- After working hours or when effective monitoring cannot be maintained, acoustic couplers will be secured to prevent use by unauthorized personnel.
- The telephone number of the computer's communication controller will be given out only to authorized users. Telephone numbers will not be posted or inappropriately revealed, and they will be changed at least semiannually, or more often if compromise is suspected.
- Where feasible, "call back" software or procedures will be employed that will require positive identification of the calling location.

4. *Protection of Time-sharing Terminals*. Remote terminal devices having access to time-sharing services will be located in rooms/areas that can be secured after normal working hours.

Environmental Security Protection

The protection of sensitive computer facilities from the effects of fire, flood, wind, and other natural phenomena is an important element in any comprehensive and cost-effective physical security program. Some computer installations require a higher level of environmental protection than others. Such factors as the monetary value of the equipment and operational requirements for uninterrupted system availability, local site considerations, and the uniqueness and resulting difficulty in replacing system components will influence the amount of resources committed to ensure that an adequate level of environmental security is provided. Protective measures will include fire, flood or water seepage, wind, earthquake, and/or other naturally caused phenomena. All these should be considered during the site selection, computer room construction, or retrofitting of the area.

Protection of Supporting Utilities

Every AIS facility depends upon various supporting utilities, including electric, air conditioners, communication circuits, water, and other essential services.

1. Buildings or rooms housing uninterruptable power service (UPS) will be accorded an appropriate level of physical and environmental protection. Any opening that can be used as a point of access shall be protected against surreptitious entry (for example, bars and/or screens over air conditioning duct openings). All sites with UPS need to be tested on a monthly basis.
2. Other electrical facilities that support the computer facility system—including peripheral support areas, such as electrical closets and transformer vaults—will be secured and checked periodically.
3. Cables, wiring, and all communications equipment associated with a teleprocessing computer system will be protected in a manner that prevents unauthorized use or access.

DESIGNATION OF AREAS

1. Each AIS facility should be designated and conspicuously posted as a controlled-access area. The term controlled-access area, as used here, is defined as any area to which access is subject to special restrictions or controls for reasons of security or the safeguarding of property or material. The degree of security and control required will depend upon the nature, sensitivity, or importance of the security interest or other matter to be protected. Different areas involve varying degrees of security interest depending upon their purpose, the nature of the work, information, and/or materials concerned. The entire area may have a uniform degree of importance requiring only one level of restriction and control. Differences in degree of importance may require further compartmentalization of certain activities. To meet these different levels, a restricted area or portions thereof may be further designated as an "exclusion," "limited," or "controlled" area. Table 12.1 identifies the AIS areas and security levels associated with each area.
    - *Exclusion Area.* All or part of a restricted area containing (1) a security interest or other matter that is of such a nature that access to it constitutes, for all practical purposes, access to such security interest or matter; or (2) a security interest or other matter of such vital importance that

Table 12.1 Minimum physical security standards for AIS site installations

	Restricted Area	Exclusion Zone	Limited Zone	Controlled Zone
AIS Center (processing all levels and specific accessed classified material)	XX			
Tape/Disc Library		XX		
Computer Room		XX		
Administrative Support Area			XX	
Support Functional Areas (electrical power, etc.)				XX

proximity resulting from access to the area is treated as the equivalent to (1) above.
- *Limited Area.* All or part of a restricted area containing a security interest or other matter; uncontrolled movement in this area will permit access to such security interest or matter, but access may be prevented within this area by escort, guard, or other internal restrictions and controls. Individuals who have legitimate reason for entering a limited area may do so if internal restrictions and controls are provided to prevent access to classified information.
- *Controlled Area.* That portion of a restricted area that is usually adjacent to or surrounding a limited or exclusion area. Entry into the controlled area is restricted to authorized personnel. However, movement of authorized personnel within this area is not necessarily controlled, because mere entry to the area does not provide access to the security interest or other matter within the exclusion or limited area. The controlled area is provided for administrative control, safety, and/or as a buffer zone for in-depth security for the exclusion or limited-access area. The degree of control of movement within this area will, therefore, be as prescribed by the physical security office and systems security office.
2. All AIS administrative areas, computer rooms, and tape/disk libraries should be protected.

ACCESS LIMITATIONS, CONTROLS, AND AUTHORITY

To provide the necessary security, the following shall be implemented:

- Positive identification is required in addition to a definitive need-to-know for entry. Visitors are under escort at all times.
- There should be a security guard, receptionist, supervisory operational personnel, duty officer, and/or electromechanical access control devices at all points of entrance and exit.
- An access listing/entry control roster will be maintained. Positive identification will be required in conjunction with the access listing/entry control roster. The number of persons authorized access will be kept to a minimum in accordance

with local operating standards, the availability of visitors' escorts, and the need to protect data within the various site installation areas.

Protective Barriers

Barriers must be a mandatory element for all areas within the site installation. A series of protective barriers should be in effect from the outermost perimeter to the innermost area of the facility. This in-depth protection is a prime necessity to assure that physical security minimum standards are met for the optimum protection of the site and data within.

1. Areas that are "caged" will have at least a 6-gauge steel mesh with two-inch diamond-style grid extending from the floor to the true ceiling. The ceiling area will also be caged with diamond grid steel mesh. All devices used to hold the steel mesh together will be spot-welded.
2. The room or building will have walls constructed of masonry or similar material extending from the floor to ceiling, with a reinforced ceiling and floor. The thickness will be at least six inches, with reinforced metal rods for exterior walls. Interior walls need only be solid-faced on both sides.
3. In all cases, the door will be constructed of at least 1 ¾ inches solid wood, which is covered on the exterior side with a sheet of 16-gauge steel. Cages will have ½-inch diameter steel bars welded to a grid with openings between the bars that do not exceed four inches in width and twelve inches in height.
4. The computer room facility area will have a minimum 1-inch covering on the interior wall consisting of ¾-inch plywood faced on both sides with ⅛-inch steel; all edges will be flush with each other and welded with a continuous weld. Such steel covering will also be located below the working area floor, at the true floor level, and at the ceiling level. These are security requirements for government installations. Though neither required nor mandated otherwise, they are very expensive to install, but should be considered for those private industry areas where the threat and/or vulnerability is high, security awareness must remain high, and the concerns over the loss or compromise of AIS data must be protected to the fullest.

Protective Lighting For all exterior areas, protective lighting outside the site installation buildings will be mandatory. Exterior entrances should be illuminated to a minimum height of ten feet on the vertical surface and eight feet on a horizontal from the entranceway.

Intrusion Detection System (IDS) IDS is required on all entrances and exits. Within various zones, individual rooms, or areas IDS devices are used (in addition to any door sensors) to protect against surreptitious entry. The IDS system is monitored from a guard post that is continually manned and the IDS system monitored on a twenty-four-hour-a-day basis. All IDS system wiring has to be in a protected environment to ensure that no tampering can occur without instant detection.

Warning Signs Warning signs will be posted at the main entranceway to all major areas that require a controlled access. In addition, various areas within the entire site installation that are further controlled will also have a warning sign posted.

Security Patrol Requirements A security patrol will be conducted on the exterior of a site installation building at least once every two hours. Within a site, a security patrol will be conducted at least once every four hours. These patrols are after normal working/nonoperational hours only. During normal working hours (namely, daytime) a random patrol will be conducted at least once during an eight-hour period.

Lock and Key Control Keys to various structures and room and/or areas within structures that have various levels of controlled-access areas and security containers within these areas will be maintained under continuous written accountability and will not be removed from the installation. The use of any form of master key system must be prohibited.

1. Keys will be maintained by the security office or a designated official (that is, guard post, twenty-four-hour-a-day basis) and will be signed for by an individual needing a key to gain access to an area.
2. Personnel who work within a given area that employs a lock and key may have an individual key, provided that it has been

signed out, a receipt processed, and the key turned in upon the individual's departure or transfer to any other office, area, or installation where they will not require further continual access to that particular lock.
3. If a key is lost, the lock cylinder will be changed immediately, all previous keys still out will be returned, and new keys issued.
4. At least once a year, individual lock cylinders should be randomly rotated or rekeyed. The particular method used will be determined by the local physical security officer.

Combination and Cypher Locks When combination locks are used for security and access control purposes, they will be of the three-wheel variety with the possibility of at least 100,000 different combinations. For government or contractor installations, combination locks will be on the current GSA-approved list for security devices. Cypher locks will be on the current GSA-approved list for security devices and have the potential for a minimum of 10,000 different combinations. For both three-position combination and cypher locks, inspection team personnel should remember the following during the inspection process:

1. No matter whether such locking devices are for government, contractor, or commercial use, they should have the combination changed when an individual who had knowledge or may have had knowledge of the combination departs; at least yearly; or as determined by the local physical security officer in consideration of protection of the data or classified defense information.
2. The locking device should be tested on a random basis, at least twice yearly, by security personnel. Random testing ensures that the combination is not one that can be easily determined (for example, 10-20-30, etc., for combination locks, and 1-2-3, 2-4-6, and the like, for cypher locks).

In addition, for some areas, a high-security hasp and lock may also be used for each entranceway that is not used on a continual basis.

13
The Security Guard Force

The security guard force is the human component of the overall physical security process for a facility. By its very nature, the security guard force systematically and selectively employs humans as barriers between a potential intruder and the materials and information being protected. The security guard force gives an alarm if a situation arises that would threaten the facility security; apprehends an unauthorized person; identifies personnel entering into, within, or leaving the facility or a portion thereof; or any combination of the above three.

DETERMINATION OF NEED

To determine the need for any form of a security guard force, all factors that bear on the security interest of the facility must be considered. The nature, location, and layout of the facility and the incorporation and effectiveness of other physical security elements must be viewed. These factors vary widely between different facilities, thus it is impossible to present exacting criteria for all.

    The need for guards is proportional to the scope of activities (including value and sensitivity of information or materials), the number of personnel, the facility size, or a combination of these factors. As the factors increase, the safeguards provided by individual personnel tend to decrease, resulting in the need for a guard force to control access to the facility or a restricted area within it.

A guard force presence is very seldom required in separate offices or small shops or laboratories that involve only a limited number of people. The security interest for a small facility usually consists only of documents or materials in such small quantity that they can be adequately safeguarded by appropriate security repositories and/or personal custody by authorized personnel who use such material and information in their work. When a guard force is desirable but prohibitive because of the expense, the operational personnel are responsible for providing the physical security to the greatest extent possible. For the private sector, note that a guard or guards may be required only to control access and protect against malicious or accidental damage within the facility. In this type of situation, guards are an essential element of insurance for production, service, or overall operations of a larger organization or agency.

Security guard force personnel fall under two general types of personnel:

1. Those persons who are assigned guard duties as a primary function, and who have met selection criteria and have been trained to take appropriate action in situations that create a hazard to security. These are full-time personnel.
2. Those persons who, by the nature of their location or as an additional duty, fulfill a security function on a part-time basis.

Regarding the above, all security guard forces can be classified as proprietary, government military, government civilian, civilian contract, and foreign contract. Each classification has its own training requirements and standards. Federal guard forces may be identified as police, having fuller powers or else limited powers. Military police must meet certain federal laws and regulations in addition to individual DOD and component requirements. Civilian contract security forces may or may not have certain police powers, but these are usually identified under state laws. Foreign contract guard forces may be authorized to perform some police functions and responsibilities, depending on the laws of the country in which they are hired.

The requirements of the facility will largely dictate the type and number of personnel required to provide the security function and the training needed to allow them to fulfill their functions. The security office must evaluate the needs of the facility and, using that as a baseline, apply the fundamentals as necessary.

Realize that there are certain characteristics inherent in human beings that tend to influence their effectiveness as a security barrier. Some of these are desirable, while others are highly undesirable.

The individual with intelligence provides a potential for a flexible and versatile security medium. Because of this intelligence, humans are theoretically capable of taking remedial action when confronted with any one of a number of situations. This ability is, of course, proportional to the individual's innate intelligence and training and experience. However, the average selected and trained security guard, when properly motivated and when operating at peak efficiency, should far surpass any animal or energy-type barrier in providing all-around security and in successfully dealing with unforeseen occurrences.

Persons with opposing ideologies are to be found within any group, so it can be expected that such a person will be among those whose job is to provide a barrier against access to facilities, materials, and information. Keep in mind that such placement may provide an enemy or intruder with an avenue of access to or within that facility. No two humans react in the same manner. In establishing and/or evaluating the use of security guards as a human barrier, security personnel must consider those human traits that may compromise defenses and minimize their effects. Various background checks, personality and psychological testing, and formalized on-the-job training can weed out any undesirable elements.

TYPES OF SECURITY GUARD FORCES

A security guard force could be a contracted guard service, a proprietary guard force, civil service employees under military or civilian direction, military personnel under military and/or civilian authority, General Services Administration guard forces or other government agencies, or Federal Protective Service police. In the majority of situations, especially for those hired directly by the government, the acceptable standards (including training and capabilities) have already been determined and will apply to all facilities. Some of the contracted guard services or proprietary-type forces meet and exceed these requirements. The military is another matter, with each service determining the specific training and use requirements of the individuals. In the case of the military, training extends to areas that would not normally be utilized within other types of guard forces.

For the majority of facilities, either a contract guard service or a proprietary guard service will be used. A number of years ago, the National Advisory Committee on Criminal Justice Standards and Goals, Law Enforcement Assistance Administration, U.S. Department of Justice, provided a list of the advantages and disadvantages of contract and proprietary guard forces (see Tables 13.1 and 13.2).

Security Guard Force Mission

The mission of the security force for a given facility is closely related to the overall mission of the security office for that facility. In part, depending upon specific mission responsibilities and authorities, the guard force will have the responsibility for the following:

1. Implement and enforce the facility system personal identification and control.
2. Observe and patrol designated perimeters, areas, facility buildings, and other activities of security interest.
3. Apprehend persons and/or vehicles gaining unauthorized access to facility security areas.
4. Perform *essential* escort duties.
5. Implement and enforce an established system of control over the removal of material and documents from security areas and/or the facility.
6. Check designated depositories, rooms, or buildings of security interest outside the normal working hours to determine that they are properly locked and secured or are otherwise in order.
7. Report to a designated supervisor—as a matter of routine under normal conditions—and in the event of unusual circumstances, as necessary.
8. Respond to protective alarm signals or other indications of suspicious activity.
9. Act as necessary in the event of situations affecting the security of the installation, including fires, accidents, internal disorders, and attempts to commit espionage, sabotage, or other criminal acts.
10. Otherwise, generally guard data, materials, and equipment against unauthorized access, loss, theft, or damage.

Table 13.1 Advantages and Disadvantages of Contract Guard Services

Advantages	Disadvantages
1. Selectivity—employer retains only those persons personally approved.	1. Turnover—extremely high throughout industry.
2. Flexibility—more or fewer personnel, as required.	2. Divided loyalties—serving two masters dilemma.
3. Absenteeism—replacement of absentees on short notice.	3. Moonlighting—low salary for guards may force them into secondary jobs, resulting in tired and nonalert personnel.
4. Supervision—supplied at no cost to the client.	4. Reassignment—some agencies send the best personnel at inception of the contract and then replace with others afterward.
5. Training—supplied at no cost to the client; may be superior to in-house training program.	
6. Objectivity—judgment not clouded by personalities.	5. Screening standards—may be inadequate.
7. Cost—20 percent less than in-house, not counting administrative savings (e.g., insurance, retirement pension, social security, medical care).	6. Insurance standards—determining liability and ensuring individual guards are bonded and insured.
8. Quality—may be of higher caliber than an in-house guard.	
9. Administration and budgeting—brunt borne by guard company.	
10. Unions—very little problem, because contract guards are usually not unionized.	
11. Variety of services and equipment—guard company may be specialists in various criminal justice skills or expensive equipment unavailable to in-house security.	
12. Hiring and screening costs—borne by guard company.	
13. Better local law enforcement contacts—may know more law enforcement personnel.	
14. Sharing expertise and knowledge—may have developed security skills as a result of many jobs; can be shared with the client.	

Table 13.2 Advantages and Disadvantages of Proprietary Guard Forces

Advantages	Disadvantages
1. Loyalty—a positive quality.	1. Unions—may go out with the company union, refuse to cross picket lines, etc.
2. Incentive—promotion possibilities within the entire company structure.	2. Familiarity—may become too familiar with personnel to be effective on the job.
3. Knowledge—of operation, products, personnel of the organization because of permanent employment.	3. Cost—expensive (salary, benefits, workmen's compensation, social security, liability, insurance, work space, equipment, training)
4. Tenure—less turnover than contract guards.	4. Flexibility—hard to replace absent personnel.
5. Control—stays inside company structure.	5. Administrative burdens—must develop an upper-level staff to handle these personnel.
6. Supervision—stays inside the company structure.	
7. Training—can be specifically geared to the job performed.	
8. Company image—may become a status symbol.	
9. Morale—a hoped-for state maintained by security manager.	
10. Courtesy—can render courtesies to VIPs because of familiarity with company personnel.	
11. Better law enforcement liaison—security manager can informally develop law enforcement liaison with less conflict.	
12. Selection—company selection procedures can apply.	
13. Better communication—more direct.	

Security Guard Force Supervision

The supervision of the guard force is necessary to assure the overall effectiveness of the various mechanics of guarding, detecting, and taking action appropriately in case of an emergency or unusual circumstances. The morale and general efficiency of the individual guard depends in large part upon the quality of supervision.

One individual should be placed in charge of each shift of guards as a shift supervisor. For smaller facilities, the guard supervisor will normally fill this position for the day shift, and subordinates will maintain it during all other shifts. A clear and definite understanding must exist as to seniority and who is in charge of the guard force personnel at any given time.

Guard personnel are assigned to shifts in accordance with local requirements, but shifts should be scheduled so that they do not coincide with any local requirements for other employee shift changes.

Guards who are assigned to a fixed post should have some designated method of securing a relief when required for short periods of time, to include lunch periods. Where fixed posts do not permit guards to move, such arrangements should be made so the guards may leave their posts for a very short break at least once every two hours.

Facilities using a guard force of several men per shift require full-time personal supervision. The ratio of supervisory personnel to guards should not exceed one supervisor to twelve guards.

The personal supervision of guard force personnel should include:

- Each guard should be inspected by an appropriate supervisor prior to reporting for duty on each shift. At such times, special instructions or orders are given, as necessary.
- Each guard post, patrol, or other activity should be *personally* contacted by the supervisor at least once per shift to determine that personnel and the post or patrol are functioning properly.
- Each guard post, patrol, or other activity should be contacted by the supervisor at least once every two hours.
- Where a guard station is fixed and only one guard is used (as is the case for very small facilities or within a controlled area of a larger facility), the site security manager may perform

the day-to-day functions listed above. In this case, a guard supervisor may only have contact with the individual guard two to three times a week, but never on a specific fixed schedule.

Security personnel at the site, in addition to the guard supervisors, should ensure that guard force personnel are not used to perform various functions that do not fall under their general or specific guard duties.

Such nonsecurity functions take away the manpower that would otherwise be utilized in the performance of security duties. When the security force has been contracted, such actions may well be outside the scope of the contract, may require extra funding, or because of the contract, may require that another guard be brought on-site. Such situations can become detrimental to the facility security posture in terms of budgeting, actual funding, and possibly certain legal implications arising from clauses of the contract.

Security Office Concerns

Within the facility security office, concerns are raised, or must be raised, considering the operations and instructions for the security guard force. On-site policies and functional instructions are usually prepared at the facility level and used by the guards in the performance of their duties. While general functional areas may be included within a guard contract, the specifics are developed and applied on-site. They include, but are not necessarily limited to:

- Company policies of the (contracted) guard force
- Policies of the company employing a guard force
- Guard post orders and instructions
- The guard post book
- Various checklists and procedures

The operational focal point for protection can be the security force working under the facility security manager. As such, the enforcement of applicable policies while maintaining a continuous protective posture for the facility is a must. The security force must be properly equipped, trained, and managed. Company policies, which security personnel are assigned to monitor, should be appropriate, enforceable, and applied consistently.

The arming of guard force personnel depends upon the needs and levels of protection required for a specific facility. When not actually required, guard personnel should not be armed. Ultimately, the effectiveness of the security force as a part of the integrated physical security program for the facility will depend on how well it is applied and maintained by security office management.

When inspecting a facility, the PSI team will be looking at the management functions as directly applied to the security force as well as the day-to-day guard functions in actual practice. It is possible, though not always necessary, to review the guard contract to determine the nature and extent of services to be performed. The PSI team should then follow up with a review of post orders, instructions, and facility policies to determine if all requirements for the satisfactory protection of the facility are being met. Observation and interviews with security guard force personnel are feasible ways to determine the training and adequacy of guard operations for the facility.

When a deficiency is detected, it may be resolved on the spot by security management or the guard supervisor. Gross deficiencies are seldom found in the guard force arena. A review of specific requirements, training, and guard usage will be most beneficial in developing an appropriate checklist for the inspection of guard force personnel and their functions for a facility.

Qualifications, Selection, and Training

A security guard force constitutes one of the most important, but singular, elements of the overall security program for a facility. The continuing cost of guarding the facility, in most cases, represents the largest continued item of security expense. Therefore, the use and deployment of the force should be carefully planned, monitored, and continuously reviewed so that the most effective and economical utilization of this manpower is obtained by security management.

The guard force requirements, utilization, and deployment for a facility must be determined after an evaluation of all factors, including the importance, classification, and vulnerability of data, materials, equipment, and the various subject areas included within the guard's training.

Tied closely to this is the background of the individuals who will be on the security guard force. All guards should be investigated and cleared for access. Minimal investigative standards are in effect at the federal and state levels, such that certain standards must be met before a guard can be hired and trained to fill specific job-related functions. These standards vary from state to state; thus, guards are licensed by the state. In some localities, depending upon state and local laws, personnel may also be "auxiliary" policemen with limited powers.

By the very nature of their duties and responsibilities, a guard may, through necessity or the inadvertence of others, gain knowledge and/or custody of vital information, government classified material, or material of a highly strategic and/or monetary value. Unquestioned loyalty and integrity is, therefore, an essential requirement for all security force guard personnel.

At a minimum, guard force personnel qualifications must include:

- Alertness—good judgment
- Confidence—a good attitude
- Motivation—loyalty and tact
- Maturity—stable within environment
- Physical fitness—good appearance
- Discipline—courteous and honest
- Literacy (high school education or equivalent; able to pass a written and oral examination of some type; and the ability to write and use the appropriate language)
- Favorable preemployment screening
- No criminal record

Training Qualifications

As previously indicated, training qualifications and requirements can vary with the type of security guard force used. The standards are usually about the same. Within the private sector, each security guard force company has its own standards; some are above and others below the federal training requirements and standards of acceptability. The selection, however, is up to the security office when they assess the functional requirements and duties of the guard force.

Subject areas that should be included within the various courses of instruction for all security guard force personnel include:

- Care and use of weapons (whether or not the guard forces will carry them).
- Common forms of espionage and sabotage activity (including industrial/economic espionage, where applicable).
- Conditions that may cause fire and explosions.
- Duties in the event of fire, explosion, natural disaster, civil disturbance, blackout, or other emergency situation.
- General and specific guard and post orders.
- Legal considerations and restrictions, including authority to arrest or detain individuals, and so forth.
- Location and operation of important steam; water; heating, ventilation, and air conditioning (HVAC); gas valves; and the main electrical switches.
- Location and use of fire protection equipment, including the main sprinkler control valve(s).
- Location and use of first-aid equipment.
- Location of hazardous materials and processes.
- Observation and description of individuals.
- Orientation of the facility, with emphasis placed on controlled, restricted, and other sensitive vulnerable areas.
- Pass and badge requirements.
- Patrol work and procedures.
- Preparation of various types of written reports.
- Proper methods of search.
- Property control.
- Radio communications.
- Types of commonly known bombs and explosives.
- Use and procedures involving the security monitoring system.
- CCTV
- Access Control System
- Intrusion Detection System
- Use of facility communication system(s).
- Visitor control and facility escorts.

14

The Facility Self-Protection Plan

Any legitimate business venture that provides a service to individuals, another business, or the government has an inherent right—not to mention responsibility—to minimize any danger to life and property that may arise. These dangerous situations may come from the effects of a bomb threat, actual bomb situation, attack by any individual or group, terrorists, unauthorized intruders, fire, sabotage, and various types of natural disasters.

Within the private sector, the facility self-protection plan is usually a part of the corporate plan, which is prepared for a specific facility or region of the company's business area. In many cases, related companies within a geographical area or at a specific site band together to form a mutual protection pact to support each other in times of crisis.

The federal government, likewise, has such plans and policies. This chapter deals primarily with the federal government. This is not to infer that such practices, methods, and procedures cannot or should not be used by the private sector, but because the government has done extensive development for emergency planning, it is used as a basis for this chapter. The General Services Administration (GSA) provides oversight and direction to all government agencies, as required by law. The GSA requires that all facilities develop a self-protection plan in accordance with the Federal Property Management Regulation, Chapter D, subpart 101-20.5, "Physical Protection"; Public Building Services (PBS) Pub 2460.1A, "Disaster Control and Civil Defense in Federal Buildings"; and PBS Pub 5930.2A, "Physical Protection." The authority of GSA for the protection of life and property in federally owned and occupied buildings, and its responsibilities for emergency preparedness programs in such buildings, is derived from US Statutes and Presidential Executive Orders.

Whether the facility is a government entity or not, *all* facilities should have a self-protection plan. Such a plan can be given a variety of names or titles, including "Facility Self-Protection," "Facility Disaster and Recovery," "Plant Emergency Operating Plan," and similar titles.

Such a plan designates the potential threats, procedures, countermeasures, training, duties, and responsibilities of assigned personnel to various functions during such dangers—or potential dangers—including warning and alarm systems, and emergency and evacuation procedures. The plan should be updated as necessary.

During a physical security inspection, consider viewing the facility emergency plan. Such plans are developed for the protection, removal, and/or destruction of material in case of a fire, natural disaster, civil disturbance, terrorist activity, or any other action that may affect the proper protection of such material. The facility plan must address and establish *detailed* procedures and responsibilities for the protection of corporate or government classified material to ensure that such material does not fall into the hands of unauthorized persons, either accidentally or by design. All facilities must have such a plan whether they are within the contiguous forty-eight states, outlying states, territories or other possessions, or within foreign countries, regardless of whether they are on U.S. military installations. Where applicable, however, an installation plan may be supplemented to meet specific facility needs.

These planning procedures do not apply to government material related to communications security (COMSEC). Planning for emergency protection and/or destruction under no notice conditions will have to be developed, when warranted and applicable to each facility, in accordance with National Security Agency instructions for such materials.

EVACUATION/DESTRUCTION OF INFORMATION AND MATERIALS

All emergency plans need to provide for the protection of information and materials in such a manner as to eliminate the risk of injury or life of facility personnel. In the case of fire or natural disaster, authorized security personnel should be placed around the affected facility, being preinstructed and trained to prevent the removal of the material. This will be an acceptable means of protecting such material and also reducing the life/casualty risk for all concerned.

All facility emergency or self-protection plans should take the following into consideration. These items, or general areas, will be noticeable in the review of the facility plan. Include only those items that apply for the facility.

1. The level and sensitivity of corporate or government-classified material held by the facility.
2. The proximity of other facilities to hostile or other potentially adversarial groups.
3. Vehicle movement schedules in the area of hostile or potential adversary groups.
4. The sensitivity of the facility operational and functional assignments.
5. The threat potential for aggressive action toward the facility. When the emergency plan for a facility is prepared, the destruction of classified or sensitive materials, to say nothing of its protection, is of prime concern. For private industry, the protection of corporate sensitive information, trade secrets, and the like, are of concern and protection may require its destruction.

The plan must also address the following considerations:

1. The reduction of the amount of material held by the facility as an initial step toward planning for emergency destruction of such materials.
2. The storage of less frequently used classified material at more secure locations within the same geographical area (if available and feasible).

3. The early transfer of as much retained corporate sensitive material as possible (to include microforms), thereby reducing the bulk that requires evacuation or destruction.
4. The determination of a specific emphasis on the priorities for destruction, the designation of personnel responsible for destruction, and the designation of places and methods for destruction.
5. The identification of the individual(s) authorized to make the final determination when emergency destruction is to begin and the means by which this determination will be communicated.
6. The authorization for the senior individual present in an assigned area containing classified material to deviate from established plans when circumstances dictate.
7. The emphasis on the importance of beginning destruction sufficiently early to preclude the loss of material. The effect of premature destruction is considered inconsequential when measured against the possibility of compromise.

The emergency plan must require that all classified and/or sensitive material holdings be assigned a priority for emergency evacuation or destruction. The priorities are based upon the potential effect upon organizational security should such material fall into the wrong hands. Holdings should be designated under one of the following:

- Priority One—Exceptionally grave damage (Top Secret material)
- Priority Two—Serious damage (Secret material)
- Priority Three—Damage (Confidential material)

If Priority One material cannot otherwise be afforded a reasonable degree of protection in a no-notice emergency situation, provisions should be made for on-site emergency destruction. For government organizations, the installation of Anti-Compromise Emergency Destruct (ACED) equipment is an option. Unfortunately, ACED equipment is not available at most facilities and is not expected to be available in the near future. This is especially true for facilities within the continental United States; ACED equipment is more likely to be found in overseas locations.

THREATS TO A FACILITY

The security posture of a facility must be concerned with both the internal and the external that is detrimental to the facility's well-being.

The internal threat concerns personnel who work in, or who have a limited or extensive knowledge of, the facility areas and the security system. This threat is generally considered to be a human reliability problem. But the level of the threat can be reduced by the incorporation of various internal security measures and procedures, such as hard-wire design, system operation, and system installation for the facility. Boxes, sensor covers, and cables should be designed to make them less vulnerable and tamper resistant.

The external threat can be divided into several categories of the skilled, semiskilled, organized, and the casual intruder. Each type of intruder would attempt to penetrate the facility barriers to conduct various operations against the facility environs. These operations can include economic or other forms of espionage, sabotage, or theft. The level of expertise, planning, types of tools or vehicles used, and the time of entry would all be considered. The use of a well-tested security system with various backups—such as IDS, CCTV, and random and roving patrols—reduces intruder potential and results in a much higher probability of early warning to the facility.

Types of Facility Threats

Fire The adequacy of fire protection equipment has an overall bearing on the basic security of a facility. It is, however, an area that receives little attention beyond ensuring that the facility has sprinklers, fire hoses, and fire extinguishers. Measures that should be included within the facility plan and reviewed throughout the facility by the inspection team members are shown in Table 14.1.

Natural Disasters Natural disasters can include flooding, earthquakes, wind storms, lightning, and the like. Such disasters take on more severe overtones when the effects of such disasters are seen. Fires, explosions, and even building collapse can be the result.

Facilities that are located in older buildings must consider the local conditions, weather history, and disasters in nearby geographical

Table 14.1 Facility threat measures that should be taken by the security/safety managers.

1. The posting and enforcement of fire prevention regulations (in compliance with any federal, state, and local laws and regulations) that are applicable.
2. A fire alarm system that reaches all areas of the facility, and is operational (check with fire department, if applicable).
3. A fire suppression system for the facility, to include hoses, extinguishers, and a secondary (backup) system(s) in effect (circumstances of the individual facilities may mean that a backup system is not, and will not, be available).
4. All fire protection equipment on-site has been properly tested and maintained (documentation should be on file).
5. Checking records and interviews with identified individuals relative to their training and stationing for personnel evacuation during a fire or other emergency.
6. Employees have been instructed and tested in the proper use of fire extinguishers.
7. Ensure that random fire drills have been conducted annually. Were they announced or unannounced? What was the minimum and maximum time required to exit building by all employees?
8. Are good standards of housekeeping being maintained?
9. See if the local fire department responsible for the facility has made a survey of the building(s) for fire safety features and discrepancies. If so, have any discrepancies been resolved?
10. Are the fire extinguishers of the right type for the type(s) of materials on hand?

areas. From these data, the facility plan will have to determine what type(s) of action will be taken when a natural disaster threat arises. No specifics are given because each type of action would depend on the facility location.

Sabotage and Espionage The threats of sabotage and espionage can be carried out by hostile or bitter current or former employees, aggressive competitor organizations, freelance entrepreneurs in economic trade secrets, or by trained agents of a foreign power. In many cases, such threats are hard to identify because of the nature of the type(s) of actions taken. Individuals engaged in such activities would try not to be seen and to work alone; when on-site, they would try to fit in.

While an espionage agent will be extremely difficult to discern, more concern can be paid toward the sabotage threat. Recently there has been an increase in violence-oriented organizations and small groups of disaffected individuals, in addition to individuals acting alone out of spite, malice, or a supposed wrong. The problem of defending the facility against the organized efforts of such individuals is related to this type of threat.

The facility should identify local threats and use any and all investigative and security resources available within the geographical area to anticipate activities that might affect the physical security of the facility and its assets.

The methods of sabotage that can be used against a facility can be generally classified as indicated in Table 14.2.

Table 14.2 Methods of sabotage to a facility generally fall within five distinct classifiable areas.

1. Mechanical - the breakage or omission of parts, substitution of improper or inferior parts, failure to lubricate or properly maintain equipment.
2. Fire - ordinary means of arson, including the use of various types of incendiary devices that can be ignited by mechanical, chemical, electrical, or electronic means.
3. Electronic or electric - interfering with or the interruption of the facility power, jamming communications, or interference with various electrical or electronic processes.
4. Explosive - damage or destruction by explosive devices or the detonation of explosive raw materials or supplies.
5. Psychological - the inciting of strikes, jurisdictional disputes, boycotts, unrest, personal animosities; inducement to create excessive spoilage or inferior work, work "slowdowns", or work stoppages by false alarms; character assassination, and the instigation of false public issues to weaken or break morale.

Table 14.3 The bomb threat reporting form should be available near each telephone

INSTRUCTIONS: Be calm. Be courteous. Listen, DO NOT interrupt the caller. COMPLETE THIS FORM.

DATE_____ TIME_____

EXACT WORDS OF PERSON PLACING CALL

_____
_____
_____
_____

TRY TO DETERMINE THE FOLLOWING [Circle as appropriate]

Caller's Identity: Male   Female   Adult   Juvenile   Age_____years

Voice: Loud   Soft   High-Pitch   Deep   Raspy   Pleasant   Intoxicated
Other_____

Accent: Local   Not Local   Foreign   Regional_____

Speech: Fast   Slow   Distinct   Distorted   Stutter   Nasal   Slurred   Lisp

Language: Excellent   Good   Fair   Poor   Foul   Other

Manner: Calm   Angry   Rational   Irrational   Coherent   Incoherent
Deliberate   Emotional   Righteous   Laughing   Intoxicated

Background Noises: Office Machines   Factory Machines   Bedlam
Trains   Music   Animals   Quiet   Voices   Mixed   Airplanes
Street Traffic   Party Atmosphere

Other_____

ADDITIONAL INFORMATION:_____
_____
_____
_____

Action taken immediately after call:
_____

_____   _____
Receiving telephone number   Person receiving call/room nbr
                                         [please print]

Bomb Threats   A bomb threat can be either a threat or an actual bomb, but in any case, it must be treated as real. Facilities must take measures to minimize the production and placement of bombs, including reducing the disruptive effects of a bomb threat.

A bomb threat can be received by a telephone, a package through the mail, or a written message through the mail. All personnel within the facility should be provided with basic information to alert them to suspicious packages and inform them concerning whom to contact. They should also be instructed in what NOT TO DO and what notes to take. Table 14.3 illustrates a bomb threat reporting format. A copy of this should be available to every individual at his or her work station, located within easy reach, usually near a telephone. Although the mail room would seem to be the likely spot for such information, it is always possible that the package is hand-delivered and may not go through the mail process.

Chapter 16 provides an overview of threat vulnerability and criticality for a facility. It shows in outline form each area that must be considered, including the identification of critical functions and, based on these, the vulnerability assessment. While this appendix may not seem relevant, it is helpful to provide a copy to the facility security manager.

Receipt of a Bomb Threat

The following procedures should be followed, when possible, whenever a bomb threat is received. The problem is that unless the individual has actually received a bomb threat call in the past, he or she is relying on memory. It is hoped that the person being interviewed has a copy of the bomb threat form near the telephone and will use it when answering any questions. Thus, a PSI team member should use the following when interviewing employees, either scheduled or randomly:

1. If a call were to be received advising that a bomb has been planted at the facility, what would the receiver do? It is hoped that the response includes to immediately signal a co-worker to listen in on an extension; remain calm and as casual as possible in conversation; engage the caller in conversation to keep him or her on the line as long as possible. Next, was the facility bomb report form annotated to include the date and time

of the call, every spoken word of the caller (if possible), and any response given? During the call, would the recipient attempt to ascertain:
- Time the bomb is set to detonate;
- Exact description and location of the bomb; and
- Caller's name and place calling from

2. During the call, it is hoped that the recipient pays close attention for any background noises such as motors running, music (including a tape), or other sounds that might provide a clue regarding the call's origin.
3. During the call, listen closely for voice quality (male or female), accents, speech impediments, and clues regarding the intellectual status of the caller.
4. The recipient—if possible—should inform the caller that the building is occupied and that the threatened detonation could result in death and serious injury to many innocent people. Note that such ploys have succeeded in keeping callers on the line and have provided additional information.
5. The recipient must attempt to stay calm and not panic. By acting calm, there is the possibility of obtaining more details from the caller.
6. If the caller indicates that the bomb is to go off "immediately" or within the next ten minutes, the recipient must immediately advise the security office that personnel should leave the area or building where the bomb is located or suspected to be located.

With a bomb threat, consider (within the facility plan) whether or not a bomb search of the facility is warranted. What about evacuation? Is there a bomb squad that would respond to a threat, and is their telephone number prominently posted or are all personnel informed of it, as necessary? If a search is conducted, will it be performed by facility personnel or others?

Threat Levels

Each facility should consider its own threat levels. Many organizations (with more than one facility) have already developed an organization threat program. The program itself may have sufficient pertinent details

Table 14.4 Sample threat level guidelines. This is only one set of threat level conditions and possible response capabilities by the facility security force. There are others, and specific threat level guidelines should be developed for a each individual facility as determined by the type and/or number of threats known and expected in a given geographical area.

Threat Level Condition	Nature of the Threat	Required Security Capabilities
Low	Standoff surveillance/ espionage.	Denial of surveillance and penetration
		Detection and deterrence of intruders
	Minimum/occasional penetration	Selective surveillance of critical areas
	Limited pilferage	Deter intruders
	Minor demonstrations	Apprehension of pilferers
		Record events (video/audio)
Medium	All of low threat intensified	Intensify response to low threat
	Sabotage	Early detection
	Harassment	Immediate response (small groups)
	Minor destruction/ disablement	Increase mobility of response force
	Dissident demonstrations	Identification and location of sabotage
		Apprehension of intruders
		Record events (video/audio)
High	All of medium threat intensified	Intensify response to medium threat
		Complete penetration denial
	Organized attack; direct conflict	Immediate response to groups
		Record events (video/audio)
	Major destruction	Apprehension of intruders
	Combat-type intelligence gathering	Sabotage detection and prevention
		Remotely controlled and/or automated response capability
		Interface with other protective forces (police, military, etc.)

to cover individual facilities. For those facilities on a larger installation, the installation threat levels should be used. Facilities themselves should consider, at minimum, three levels of threat and determine the nature of the threat and the required capabilities of the security force for that facility.

The essential point is that each threat level is an escalation of a previous one, with each higher threat level requiring a certain number of increases in security posture. The facility security plan—and, of course, the facility self-protection plan—must take this into effect, detailing each configuration to meet the existing threat while still providing the required amount of interface capability to upgrade with minimum difficulty and expense.

Table 14.4 provides a breakdown of threat level guidelines in terms of the appropriate levels of response by a local security force. Note that as each level of threat rises, the requirements for the probability of threat detection, reliability, degree of security required, and the speed and intensity of the security force also rises.

15
Asset Protection and Loss Control

In terms of property control and loss prevention, the prevention of thefts from offices, shops, general storage, high-value areas, and warehouse areas must be addressed. These areas are ripe for such thefts, especially for items that can be carried out by hand and are of a high value, allowing for a quick resale on the outside. This chapter provides discussion of areas in which problems arise, along with typical solutions. Security personnel must bear in mind that the development of the program starts with an awareness that such problems exist, and that they concern all personnel within the facility, visitors, and maintenance personnel who work in and around the facility.

Managers at all levels must set the example for property protection and loss control. All facility employees should know that the security office *and* senior management care about loss prevention. This should be stressed through the weekly bulletin, flyers, and posters in the facility. Company policy must also set forth standards of acceptability and nonacceptability with regard to property retention, control, and loss prevention.

OFFICES

Situations in which offices are temporarily left unattended create the opportunity for individuals (visitors, workmen, other facility employees) to walk in and remove items. Visitors need to be monitored and escorted at all times; they should never be left alone in an office. Unknown persons found walking around or in areas in which they do not belong or who are "lost" should be challenged. The security office should be contacted and/or the individual personally escorted to the security/visitor area. If visitors have packages or briefcases, such items should be inspected or retained at the security office until the personnel leave the facility whenever possible.

Conference rooms adjacent to office areas should be kept locked when not in use, *and* items of value should either be secured within a conference anteroom or removed to another area and secured.

Displays for public viewing can be within main hallways or just outside in hallways or general waiting areas. Some displays for public viewing or internal employee use and education have items of value in them. It takes only a moment to have a display case "accidentally" broken and a small item removed and concealed under the guise of holding an injured arm or hand to the body. Most personnel tend to think of the injury and not the items within the display case. When displays are removed for storage, they should be stored immediately and not left sitting to the side or tucked away near an empty corner. Empty display cases can block from view or conceal what is on the other side of them, including hidden personnel, stolen equipment, or even a potential entry point that has been made or used in the past for access to another area.

PERSONAL PROPERTY

Familiarity among employees tends to create a relaxed atmosphere. This develops into a lack of direct concern for personal property that is not of a high value. It is incorrect to assume that the office is secure because only facility employees use it. Employees should be encouraged to leave nonessential items at home, especially those with a high resale value. Lockers can be provided for the employees, allowing them to use personal locks. Warning notices should be posted in all lavatories reminding personnel of the danger of leaving items of value in the washroom areas, on desks, and so forth. Handbags, purses, briefcases, watches, rings, other jewelry, cash, and other valuables should *never* be left unattended. At the minimum, these items should be locked in desks or cabinets and the key retained by the individual.

KEY SECURITY

Keys should never be labeled to indicate the office or room they open. Keys should have a code number only imprinted. They may indicate "property of" for large facilities, but never for facilities that are small. Master keys or similar passkeys should never be labeled to indicate that they may open more than one lock, nor should they indicate a specific facility where they are used. A strict system of key security and accountability procedures must be developed, used, *and* enforced. Master keys must have a very limited distribution within the organization.

CLEANING CREWS

Maintenance and janitorial staff or contracted cleaning companies have a high turnover of personnel. Limit the use of any passkeys; never issue a master key, but only keys to local offices. When possible, have a security force guard accompany cleaning personnel (small facility) or check on personnel hourly (on a random basis). Cleaning crew personnel should arrive and depart by the same door, bringing in only their lunch. Their work supplies and equipment should be stored on site. Any personal belongings should be secured in a locker before they start their work.

It is best if the doors to the areas the staff are cleaning are left unlocked. These should be locked immediately afterward, or when they move to a new area of a facility building. A security guard should ensure the doors are properly locked. This should not be done by other maintenance personnel. All cleaning and janitorial staff personnel should have proper badges with picture identification. New personnel should be processed through security and be given an identification badge before they start to work. Never accept a "new" employee who is "filling in" for somebody else who is sick that day; in other words, "know the crew."

WORKMEN

Check the identity and status of all workmen at the facility. Develop procedures to register and identify them by name, their company, where they will be working within the facility, time in and out, whether or not an escort will be required, the name of the escort, their office location, and telephone number. If the workman is not a regular who works full-time on the site, he should never be left alone, especially in areas where there is sensitive facility information, classified documentation, pilferable equipment, and high-value items.

SHOPS, STORAGE, AND WAREHOUSE HOLDING AREAS FOR SUPPLIES AND EQUIPMENT

All facilities have some supplies and equipment that are maintained in storage or work areas. For these areas, consider the dishonest employee. Don't entrust cash or large-value hand-carried items to new employees. If possible, consider employee bonding where appropriate.

False or duplicate keys to such areas can be a problem. For this reason, use high-security locking devices. Key control is a must. Limit the availability of keys to such areas to those individuals who actually are required to lock and unlock the areas on a daily basis.

Consider the possibility of entry from a ceiling, adjoining building, or through a wall, vent, crawl space, or basement area, to say nothing of entry from closed-off passageways. Ensure that such areas have alarm devices or adequate lighting.

When warehouses or storage areas are being built, meet with the architect early on and incorporate security features before construction. This is the time to eliminate some potential vulnerability points.

Asset protection and loss prevention are additional tools that are used by the security staff in all organizations, no matter what their size. It behooves all security individuals to at least be well-read in these areas. Without a background and knowledge of the subject matter, inspection team personnel may consider only items that prevent problems from the outside; the majority of loss prevention incidents come from insider threat to a facility and its assets.

16
Facility Threat, Vulnerability, and Criticality

The priorities for protection and level of threat vulnerability and facility criticality are, most definitely, subject to change. Functional and environmental changes to a facility and geographical area will mean changes within the criticality and vulnerability to the facility functions and work environment.

By definition, criticality measures how essential the function(s) are to the routine of a facility (and restoration of facility operations, if necessary). Vulnerability measures how susceptible the functions are to damage, destruction, or loss resulting from hazards created by natural or human forces.

Losses to an organization can be of a physical, intellectual, data (such as corporate data or information), or of a monetary nature.

During a PSI, a function that is studied and taken into consideration as part of the facility operations becomes a possible critical function. In the absence of that function or functions, the capabilities of the facility will suffer in some form or manner. By looking at all the functions, certain vulnerabilities may emerge. Each vulnerability that is

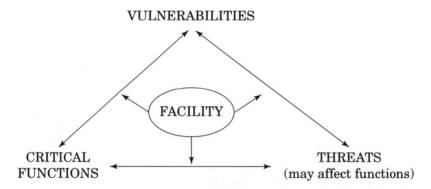

Figure 16.1 The interrelationship of threats, vulnerabilities, and critical functions must be understood and addressed carefully when conducting the facility inspection, thus ensuring that proper protective measures have been identified and implemented for the facility.

identified should be tied to all of the functions it may impact should it be exploited. The point to remember here is that this type of information is usually absent from an inspection checklist. It must be determined during a physical inspection of the facility.

Below is an outline that can assist during a PSI to determine the facility's critical and vulnerable points. Figure 16.1 illustrates the interrelationship of these functions, hazards, and vulnerabilities to a facility.

A. Identification of critical functions.
 1. Importance of functions (what is important to facility planning, growth, survivability?)
 2. Factors that can affect time loss (restoration of services)
    a. Short supply
    b. Long lead time
    c. Time to substitute
 3. Functional area criticality study
    a. Group A—fatal to the facility
    b. Group B—very serious to the facility
    c. Group C—moderately serious to the facility
    d. Group D—relatively unimportant to the facility
    e. Group E—unable to assess at this time

B. Vulnerability assessment.
   1. Hazards (to be identified according to the function they affect)
      a. Natural
      b. Human caused
   2. Vulnerability factors
      a. Security force
      b. Physical security
      c. Location of critical functions
      d. Attended or unattended
      e. Supervision (or lack thereof)
      f. Target analysis
         (1) Physical
         (2) Social
         (3) Political
         (4) Economic
         (5) Services
      g. Access to critical functions
C. Functional and environmental changes can produce further changes in the criticality and vulnerability of a facility. Redundancy (or lack thereof) becomes the primary factor for self-protection and sustainment of the facility. Some examples are:
   1. Changes in equipment
      a. New equipment
      b. Equipment replaced by personnel
      c. Personnel replaced by equipment
      d. Relocation
   2. Changes or modification to products or processes within the facility
   3. Changes in fuel source for facility buildings
   4. Changes in location or configuration of the facility
   5. Changes in power supplies for the facility
   6. Changes in transportation system to/from/around the facility
D. Countermeasures
   1. Reduction of vulnerability
      a. Physical security measures
      b. Inventory, audit, and/or accountability programs
      c. Good housekeeping
      d. Security force

    e. Warning signs
    f. Employee education
    g. Supervisory responsibility
    h. Report within community; liaison with critical counterparts within surrounding community and other facilities within location area
    i. Relocation functions; dispersal functions
    j. Rerouting of production traffic
    k. Move
    l. Runoff ditches, blast walls, and other facility structural strengths
    m. Direct attention away from facility
    n. Decoys, deception, and camouflage
    o. Fire fighting capabilities and limitations within and around the facility
    p. Maintenance program
    q. Inspection program
  2. Reduction of criticality (examples)
    a. Alternate source of supply, rapid delivery, stockpile, subcontract, alternate methods
    b. Latest products
    c. Salvage
    d. Backup power and parts
    e. Duplicate records and files
    f. Mutual aid
E. Survey Officer Mission
  1. Urge management to identify and document critical functions and assess vulnerabilities
  2. Identify additional critical functions
  3. Identify additional vulnerabilities
  4. Recommend countermeasures

Criticality and vulnerability importance must not be underestimated. The difficulty of replacing a function or the continuation of such function is such that the loss, disruption, or impairment of the function would adversely impact the capability to continue operations necessary to the facility.

First, the functions must be identified in relationship to the overall facility operational needs. Second, the determination of the threat to facility operations and the likelihood of its occurrence must be made. Finally, an assessment is made of the vulnerabilities of critical functions to the threats.

These are the essential preliminaries on which the formulation of protective measures must be based. Such measures are designed to reduce criticality and vulnerability factors to a manageable level. All vulnerabilities cannot be totally eliminated. It is essential, then, to keep in mind that criticality, vulnerability, and threats are relative to the operational environment in which they are found. Therefore, the survey process must be an ongoing effort to keep attuned to an ever-changing environment.

Appendix 1
Security Inspection Checklists

The purpose of this checklist is to provide physical security inspection team personnel with guidelines for evaluating existing security measures. It is not intended to be all-inclusive. The original copy will be maintained by the PSI team with a copy provided to the facility manager.

The checklist has been divided into various sections, by subject, to allow the user easier reference. Copy the checklist, as necessary, extracting those portions required for your physical security inspection. Add or delete items to ensure that the checklist is as specific as possible. Before using the checklist, make an extra file copy.

As a site-specific checklist is developed, a "headline" should appear on the first page as shown in Table A.1. This provides an identification of which area(s) of the facility on the checklist were inspected, when they were inspected, and how long since the last inspection.

Facility Security Office

1. Is a security manager designated in writing for the facility?
2. Are various level security officers (division, branch, and so forth) designated in writing?

Table A.1 Sample "headline" at the top of the first page of the inspection checklist. It provides the basic information about what facility was inspected and the date(s) of inspection.

                                      (current date)
ACTIVITY/FACILITY BEING INSPECTED:_____
INSPECTION DATE(S):_____
LAST INSPECTION OF FACILITY WAS: _____

3. Are the security manager and security officers aware of their various duties, functions, and responsibilities within the security arena?
4. Does the facility/activity have a current security plan? What is the date of the plan:_____
5. Does the facility security office maintain a current listing of security regulations and references to include? (list appropriate references here)
   a. DOD Reg 5200.1-R
   b. Industrial Security Manual
   c. Industrial Security Regulation
   d. Other:
6. Does the security plan contain annexes to cover:
   a. Fire
   b. Explosion
   c. Crises/disturbances
   d. Terrorist threat
   e. Sabotage
   f. Bomb incidents
   g. Natural disasters
7. Within the bomb incident/threat portion, does the plan include:
   a. Preventative measures to reduce the opportunity to introduce a bomb into the area?
   b. Procedures for evaluating or handling threatening messages?
   c. Policy for evacuation and safety of personnel?
   d. Procedures to search for a bomb?
   e. Procedures for assistance (law enforcement/bomb disposal squad)?

f. Procedures for assistance if a bomb or suspected bomb is found on premises?
   g. Procedures to be taken in the event of a bomb explosion?
8. Does the guard force have a current listing of security clearances tied to organizational personnel for clearance verification purposes? If so, when was the list reviewed and/or updated?
9. Is there a package inspection program in effect for the facility? Is it required for all packages, or is it performed on a random basis? When was the last instance of its being used? During the past twelve months, what was the frequency of package inspections?
10. Does the facility/activity maintain a viable security program that includes subversion and espionage?
11. Do all members assigned to the facility/activity (contractor, government, or military) receive briefings/updates?
12. Does the security officer/manager conduct a physical security survey at least annually?
13. Have areas been designated as exclusion, limited, or controlled, as necessary, to safeguard information/property?
14. Are basic security measures for exclusion areas in effect?
15. Are basic security measures for limited areas in effect?
16. Are basic security measures for controlled areas in effect?
17. Are any of these three areas lacking in any security measures that should be put into effect (even through an area meets basic security measures criteria)?
18. Are security measures in effect to protect:
   a. Electrical power supplies/transmission facilities?
   b. Communication centers/equipment?
   c. Small arms and ammunition of the guard force?
19. Has a lock and key custodian been appointed in writing?
20. Does the lock and key program include:
   a. A key control register?
   b. An inventory of keys with each change of custody?
   c. Control of master key(s)?
   d. Control over duplication equipment and key blanks? Or are keys marked "do not duplicate?"
21. Are burn bag storage holding areas either:
   a. Locked up and under constant guard surveillance?
   b. Or locked in a secured, alarmed room?

22. When removed for transport to a destruction site, are burn bags escorted by at least two properly cleared individuals?
23. Are all portions of reproduction areas checked daily for unsecured classified documents (whether or not machine room(s) are authorized for reproduction of classified material)?
24. Are limitations put on trash separation by maintenance personnel?
25. How is normal, unclassified trash disposed of?
26. Are security container combinations changed yearly or more frequently if required?
27. Are security container combinations maintained in the security area?
28. Is the security force strength and composition commensurate with the degree of security protection required for the activity/facility?
29. Do security force personnel have clearances equivalent to highest degree of security classification of documents, material, and the like to which access *may* be required?
30. Are security force personnel inspected and briefed by a supervisor prior to reporting to their posts for duty?
31. Do supervisors inspect each post/patrol/activity at least once each shift? At least three times weekly? Is this on a random or scheduled basis?

Fire Prevention

1. Does the facility have a current fire prevention plan?
2. When was the plan last updated?
3. Is there a fire department within the facility confines or less than ten minutes' distance?
4. What is the response time for the fire department?
5. Are the parts of the fire plans that concern evacuation procedures prominently posted?
6. Are all fire exits prominently posted?
7. Are they kept clear?
8. Are the hallways kept clear?

9. Is there a buildup of combustible trash within or close to the facility buildings?
10. Are all personnel aware of fire hazards within their facility buildings?
11. Is there a fire alarm system for the facility?
12. Are fire panel alarms located in all hallways and public gathering areas?
13. When was the last time the fire alarm system was tested?
14. Are regular fire preventions made to detect hazards?
15. What frequency?
16. Are any hazards found during inspections recurring?
17. What has been done to eliminate them?
18. What first aid and firefighting equipment is on hand in terms of hoses, extinguishers, pumps, and the like?
19. What types of fire extinguishers are available?
20. Are fire extinguisher locations known to personnel who work within and near offices surrounding each extinguisher?
21. Do personnel know how to operate the extinguishers properly?
22. When was last time the extinguishers were inspected and/or tested?
23. What is the frequency of fire drills for the facility?
24. When last tested?
25. What problems were encountered?
26. Do firefighting personnel know the locations of the main sprinkler valve controls?
27. Are valve controls, pipe, or sprinkler heads in good or poor condition?
28. How often are they inspected?
29. Are fire hydrants accessible and in good working order?
30. Is there a no smoking policy for certain areas where combustibles are stored?
31. Is it rigidly enforced?
32. Are containers of explosive gases or materials stored?
33. Are the areas prominently marked?
34. Is there a photography lab on the premises?
35. What is the level of fire susceptibility to the area?
36. Have fire hoses been inspected within the past year to ensure they are not rotted and do not have holes in them?
37. Have they been tested within the past year?

Identification and Badges

1. Is a pass or badge identification system used to identify all personnel within the confines of security areas?
2. Does the identification medium provide the desired degree of security?
3. Do written instructions include arrangements for the following:
   a. Protection of coded or printed components of badges/passes?
   b. Designation of the various areas requiring special control measures?
   c. Controlled issue of identification media?
   d. Description of the various identification media involved and the authorization and limitations placed upon the bearer?
   e. Mechanics of identification for entering and leaving each area, as applied to both employees and visitors?
   f. Details of where, when, and how badges shall be worn?
   g. Procedures to be followed in case of loss or damage to identification media?
   h. Procedures for recovery and invalidation?
4. Are personnel who require infrequent access to a security area or who have not been issued a permanent identification pass or badge for such treated as "visitors" and issued a visitor's badge or pass?
5. Is supervision of the personnel identification and control system adequate at all levels?
6. Are badges and serial numbers recorded and controlled by rigid accountability procedures?
7. Are lost badges replaced with badges bearing different serial numbers?
8. Have procedures been established that provide for issuance of temporary badges for individuals who have forgotten their permanent badges?
9. Are temporary badges used?
10. Are lists of lost badges posted at guard control points?
11. Are badges of such design and appearance as to enable guards and/or other personnel to recognize quickly and positively the authorizations and limitations applicable to the bearers?

12. Do procedures exist that ensure the return of identification badges upon termination of employment or assignment?
13. Are special badges issued to contractor employees working within security areas?
14. Are all phases of the personnel identification and control system under supervision and control of the security officer?
15. Have effective visitor escort procedures been established?
16. Are visitors properly escorted within security areas?
17. What controls are employed regarding visitor movements in security areas?
18. Are visitors required to conspicuously display identification media at all times within security areas?
19. When visitors leave security areas, are they required to turn in identification badges and is the departure time in each case recorded?
20. Are visitors who indicate an intention to return at a later time permitted to retain identification badges?
21. Are guards or other persons stationed at different areas of security importance to ensure visitor control?
22. Are permanent records of visits maintained? If so, by whom? For how long?
23. Are special visitors, for example, vendors, tradesmen, utility servicemen, special equipment servicemen, and so forth issued a special or distinctive visitor badge?
24. What measures are employed, other than the issuance of identification badges, to control the movement of contractor personnel working within security areas?
25. Does the facility use a pass/badge exchange system?
26. If so, do both bear identical photographs of the bearer?
27. Are both printed from the same negative/taken at same time?
28. Does the guard compare pass/badge to bearer and also to pass/badge being exchanged?
29. Must the pass/badge be exchanged each and every time an individual enters and departs the facility during the same day?
30. Is there an accounting for each badge/pass?
31. Is there an inventory by security personnel at the start and completion of each tour of duty?
32. What type of protection is afforded visitor badges or passes when not in use?

33. Is there a location check of personnel who have not checked out of the facility/area at the close of daily operations?
34. How are visitors to the facility identified?
35. Through which entrances and through what number of entrances are visitors permitted to enter? How is this enforced?
36. Are visitors required to enter *and exit* through the same entrance?
37. Are signs containing instructions for visitors posted at facility entrances?
38. Is an escort system utilized for visitors?
39. If so, is an escort pool utilized?
40. How many personnel are assigned to the escort pool?
41. Is this adequate in terms of numbers of visitors to the facility?
42. How are escorts assigned to the visitors?
43. Are escort personnel permanently assigned?
44. Are escorts aware of areas into which visitors are not allowed?
45. How are escort personnel identified?
46. Does the escort remain with the visitor during entire period of the visit?
47. If not, what other control measures are employed?
48. What action is taken if the escort and visitor are separated?
49. Are escorts provided by the sections or persons to be visited?
50. If so, how are visited personnel notified of visitors?
51. How does the guard determine that the person desiring entry is an authorized visitor?
52. Is a record maintained to reflect both the visitor and the escort?
53. How long is the record maintained?
54. Does the escort accompany the visitor back to the entrance when the visit is completed? If not, what measures are taken?
55. What action is taken if the visitor is separated from an escort?
56. Are special visitor badges used for VIPs?
57. Are special or distinctive visitor badges used for vendors, utility, or other servicing personnel?
58. Are the visitor controls the same as for any other visitor? If not, what procedures are used?
59. Do permanent employee badges have an expiration date? If not, why not?

60. Are permanent employee badge pictures checked to determine if there has been a major change in physical appearance? If so, is a new badge required? If not, why not?
61. What deterrents to alteration or reproduction have been incorporated into the facility badge?
   NOTE: Such alteration or reproduction deterrents may combine any of the following: distinctive or intricate background design; inks or dyes on some part of the device that would be noticeably affected by heat or erasure, which would occur during attempted relamination or alteration; serial numbering; validation by official (not "stamped" signatures); secret characteristics known only to certain officials; watermarked insert paper; threads, logos or wires superimposed over the photo; ink that bleeds or fades when exposed to solvents needed to dissolve the plastic on laminated badges; fluorescent inks visible only under ultraviolet light; seal or signature appearing partially on the photo and partially on the badge itself.
62. Is any coding system incorporated into the badge to designate areas or gates of access, shifts, and so forth?
63. Are any electronic or automatic identification systems utilized?
64. If so, are badges spot-checked to ensure that the badge belongs to an authorized bearer?

Intrusion Detection Systems

1. Does the activity employ an Intrusion Detection System (IDS) or systems?
2. Does the IDS, where required or used, meet the following minimum requirements:
   a. Are balanced magnetic switches installed on all perimeter doors?
   b. Are sensors attached to window and vent gates?
   c. Are sensors attached to structural sections that do not provide penetration resistance roughly equivalent to that required for the basic structure?
   d. Are IDS signals monitored at one central point and is the guard force response initiated from that point?

  e. Are all sensor equipment, doors, drawers, and removable panels secured with key locks or screws and equipped with tamper switches?
  f. Have power supplies been protected against overload by fuses or circuit breakers?
  g. Have power supplies been protected against transient power surges?
  h. Have safety hazards been identified and controlled to preclude personnel exposure?
  i. Do IDS components meet electromagnetic interference/electromagnetic compatibility requirments?
  j. Do IDS components meet the requirements for spurious radiation set by the Federal Communications Commission?
3. Is the system backed up by properly trained security alert teams?
4. Is the alarm system for active areas or structures placed in access mode during normal working hours?
5. Is the system tested prior to activation?
6. Is the system inspected regularly?
7. Is the system weatherproof?
8. Is the alternate or independent source of power available in the event of power failure?
9. Is the emergency power source designed to cut in and operate automatically when AC power is disrupted?
10. Is the IDS system properly maintained by trained and properly cleared personnel?
11. What frequency of maintenance and inspection is required under contract?
12. What is the actual frequency of maintenance and inspection?
13. Are emergency maintenance personnel designated?
14. Are frequent tests conducted to determine the adequacy and promptness of response to alarm signals?
15. Are records kept of all alarm signals received, including time, date, location, action taken, and cause for alarm? How long are records kept?
16. Are supervisory personnel reviewing alarm records?
17. How frequent are nuisance alarms and what action is taken to reduce their number?

Key Control Systems

1. Has a key control officer been appointed?
2. Are all locks and keys to all buildings and entrances supervised and controlled?
3. Are keys issued only to authorized persons with a valid need for them?
4. Is the removal of certain selected or all keys from the premises prohibited?
5. What is the basis for the issuance of keys?
6. Is an adequate log maintained of all keys that are issued?
7. Are keys turned in during periods of absence (leave, vacation, and so forth)?
8. Are keys turned in when an employee transfers to another office or leaves the facility?
9. Are locks changed when a breach of security occurs to a locking device?
10. Are keys audited at least annually, with a requirement to actually produce the keys for a visual inspection?
11. Are employees authorized to duplicate keys?
12. Are keys duplicated within the facility by security personnel?
13. If so, who duplicates the keys? Where, when, and how are they duplicated?
14. Are duplicated keys afforded the same protection level as the original key?
15. Is there a key cabinet or locked container for all keys?
16. Are master keys and core keys maintained in a security container separate from other keys? (Master and core keys are never stored in the same container.)
17. If a building underwent construction or substantial changes, did the contractor have access to individual or master keys? Were any retained?
18. Are keys of the type that local access to blanks is difficult?
19. Are key blanks only available from the lock manufacturer and not through various local supply sources where keys are made?
20. Are current records maintained for the number and identification of all keys that have been issued?
21. Do records indicate which buildings or entrances the keys operate?

22. Is there a series of pre-keyed locks maintained, or pre-keyed codes, within the security office?
23. What level of protection is afforded them?
24. Are certain master and/or submaster keys, or individual room or area keys maintained by the guard supervisor?
25. Are such keys maintained within a locked security container?
26. Who else has access to the container besides the guard supervisor?
27. Do certain keys that are frequently checked out on a daily basis stay under the supervisor's control or are they maintained at a guard post?
28. Are all keys that are under control of guards inventoried daily and at shift changes?
29. Are losses or theft of keys promptly reported and investigated, as necessary?
30. Are locks immediately changed to preclude use of a lost or stolen key?
31. Are lock keys for inactive gates maintained in a security container by the guard supervisor?
32. Are padlocks rotated annually? Semiannually?
33. Are all door locks, keyed padlocks, and combination padlocks numbered for easy control?
34. Is the manufacturer's serial number or code on a lock obliterated?
35. Are measures in effect to prevent the unauthorized removal of locks on open gates, fences, and the like?
36. Are highly pick-resistant key cylinders used on sensitive, controlled, restricted, or high-value areas?

Lighting

1. Is the perimeter of the installation and security area fencing provided with adequate lighting?
2. Does the protective lighting meet adequate intensity requirements?
3. Are the cones of illumination from lamps directed downward and away from guard personnel?

4. Are lights mounted to provide a strip of light both inside and outside the fence, where appropriate?
5. Is perimeter lighting utilized so that guards can remain in comparative darkness when necessary for observation purposes?
6. Are lights checked for proper operation prior to darkness?
7. Are repairs to lights and the replacement of inoperative lamps performed immediately?
8. Is additional lighting provided at active portals and points of possible intrusion?
9. Are gate guard houses provided with proper illumination?
10. Are lighting power supplies protected? How are they protected?
11. Does the activity have a dependable source of power for its lighting system?
12. Does the activity have a dependable auxiliary source of power?
13. Is the protective lighting system independent of the activity lighting or power supply system?
14. Is there provision for standby or emergency protective lighting?
15. Can the emergency lighting equipment be rapidly switched into operation when needed?
16. Is the standby or emergency lighting equipment tested? How frequently? During the last test, did all emergency lighting equipment respond quickly? If not, why not?
17. Is wiring tested and inspected periodically to ensure proper operation?
18. Is parallel circuitry used in the wiring?
19. Is the protective lighting system designed and locations recorded so that repairs can be made rapidly in an emergency?
20. Are multiple circuits used?
21. If so, are proper switching arrangements provided?
22. Is a closed loop used in multiple circuits?
23. Is wiring for protective lighting properly run?
24. Is it in tamper-resistant conduit?
25. Is the wiring installed above or below ground?
26. If above ground, is it high enough to preclude the possibility of tampering?
27. Are switches and controls properly located, controlled, and protected?
28. Are they weatherproof and tamper resistant?

29. Are the controls readily accessible to security personnel?
30. Are they located so that they are inaccessible from outside the perimeter barrier?
31. Is there a centrally located switch to control protective lighting?
32. Are emergency maintenance personnel designated?
33. Is there adequate lighting for guards on indoor routes?
34. Are materials and equipment in shipping and storage areas properly arranged to provide adequate lighting?
35. Is there an emergency lighting system or are there individually located systems *within* the facility, especially at entrances and inside sensitive or restricted areas?
36. Are they tested at least monthly?
37. Is the emergency lighting system battery operated with a continually charged (trickle-charged) battery?
38. If not, when was the battery level last checked?
39. Are the emergency lighting system(s) adequate?
40. Should the lighting system be enlarged to cover other areas?
41. If so, what areas?
42. Is any form of screening used to protect outside lighting instruments from rocks or other objects that could be thrown?
43. Do the light beams overlap so that there are no "blind" spots or areas that are in continual darkness or at a reduced lighting level?

Perimeter Security

1. Are the facility perimeter and all other security access controlled areas defined by a fence or other physical barrier?
2. State the type, height, and general characteristics of the physical barrier and obtain a photograph for exhibit purposes.
3. Are perimeter physical barriers damaged or deteriorated?
4. Is there more than one level of perimeter barriers? If so, include diagram to demonstrate each, with their parameters and limitations.
5. If a fence is utilized as the perimeter barrier, does it meet the minimum specifications for security fencing?
    a. Is it of chain-link composition?
    b. Is it constructed of 9-gauge or heavier wire?

     c. Is the mesh opening no larger than two inches?
     d. Is selvage twisted and barbed at top and bottom?
     e. Is the bottom of the fence within two inches of solid ground?
     f. Is the top guard strung with barbed wire and angled outward and upward at a forty-five degree angle?
     g. Is the fence the proper height (including outrigger, if appropriate) in all areas?
6. If masonry wall is used, does it meet minimum specifications for security fencing?
7. If building walls, floors and roofs form a part of the barrier, do they provide security equivalent to that of a chain-link fence?
8. Are all openings properly secured?
9. If a building forms a part of the barrier, does it present a potential penetration hazard at the point of juncture with the perimeter fence?
10. If so, is the fence height increased substantially at the point of juncture? Are all portals in the barriers guarded or secured when not in use?
11. List the number, location, physical characteristics, and any deficiencies noted in barrier entrances and barrier line (use a walk-around inspection to determine this).
12. Do gates and/or other entrances in the barriers exceed the number required for safe and efficient operation?
13. Are all barrier portals equipped with secure locking devices?
14. Do all gates provide protection equivalent to that of the barrier of which they are a part?
15. Are barrier gates and/or other entrances that are not in active use locked and frequently inspected by guards or other personnel?
16. Are locks to these gates rotated annually? When was the last rotation?
17. Is the security officer responsible for the security of keys to the barrier entrances? If not, who is?
18. Are keys to barrier entrances issued to other than installation personnel?
19. Has any barrier gate or entrance been activated or deactivated since the last survey? If so, provide details.

20. Are all normally used pedestrian and vehicular gates and other entrances effectively and adequately lighted to ensure:
    a. Proper identification of individuals and examination of identification badges?
    b. The observation of the interior of vehicles?
    c. That glare from luminaries does not interfere with the guard's vision?
    d. That the appropriate signs setting forth the procedures for authorized entry (conspicuously posted at all entrances) can be seen?
21. Are "No Trespassing" signs posted on or adjacent to barriers at such intervals that at least one sign is visible at any approach to the barrier during daylight hours?
22. Are clear zones maintained on both sides of the barrier?
23. If clear zone requirements cannot be met, what additional security measures have been implemented?
24. Are automobiles permitted to park right up against perimeter barriers?
25. Do guards patrol perimeter areas?
26. What is the frequency of perimeter barrier checks made by security personnel?
27. Are barrier deficiencies immediately acted upon and are the necessary repairs performed in a timely manner?
28. Are any of the perimeters protected by an intrusion detection system?
29. Does any relocated function, newly designated security area, physical expansion, or other factor indicate the need for additional barriers or perimeter lighting?
30. If so, what action has been taken?

Building Perimeters

1. Does the main building exterior have a light-colored paint for eight to ten feet from the bottom edge to preclude intruders from hiding in shadows?
2. Are buildings clear of shrubbery where an intruder could hide?
3. Are trees located far enough away to prevent an intruder from using either height or branches as a means of entry?
4. Are lower windows locked/barred?

5. Are second and third floor windows locked/barred?
6. Are first and second floor windows alarmed?
7. Are air conditioning units in lower windows properly secured so they cannot be removed from the exterior?
8. Are perimeter areas of building(s) well lighted and free of trash buildup?
9. Are ladders, long boards, and the like, near the building or stored in a locked area?
10. Are areas away from entrances well lighted?
11. Are overhangs lighted?
12. Are outbuildings secured when not in use?
13. Are outbuildings patrolled during nonduty hours by the guard force?
14. What is the frequency of patrols?
15. Are guard patrols at random or in a set pattern?
16. Are doors and windows locked and/or alarmed?
17. Are doors made so they open outward or inward?
18. Are door pins, for outward opening doors, nonremovable or welded in place?
19. Can outbuildings be accessed without the guard force being aware of it, or in a manner that precludes casual observation of such access?
20. Are outbuildings protected by a local alarm system or tied into the main system?
21. Are plans in effect or contemplated to install monitor controls for outbuildings?
22. Are door locks rekeyed at least once every other year?
23. Are keys to outbuildings limited to essential personnel only?
24. How many keys to outbuildings have been lost or are unaccounted for?
25. Can outbuildings be accessed from another area, unobserved by a patrolling guard (enclosures against the building, and so forth)?
26. Are solid core doors used on all exterior entrances/exits?
27. Are paneled or hollow core doors lined with a metal sheet?
28. Are glass doors also alarmed or are they located at entrances that are guarded or under observation at all times, especially during nonworking hours?

29. Are unguarded glass doors protected by steel bars or wire mesh, or are they tied to the alarm system?
30. Do all glass doors have a double-keyed cylinder instead of an interior thumb-turn?
31. Are overhead doors locked either by electric power or slide bolts?
32. Are slide bolts mounted on interior or exterior sides?
33. If mounted exteriorly, are pick-resistant locks employed?
34. Do all exterior doors have a pick-resistant lock cylinder?
35. Do all exterior doors have a deadbolt feature?
36. Are all exterior doors locked and checked at the close of business each day?

Security Guard Force

1. Have security personnel been trained in procedures for the implementation of emergency and disaster plans for their activities?
2. Are all security personnel who are required to be armed also qualified with their assigned weapons?
3. Do all security personnel who are armed requalify annually?
4. Have all security personnel received instruction in the use of deadly force?
5. Are guard force weapons readily available in the event of an emergency?
6. Does the activity maintain a crisis response team?
7. Are members designated for each guard shift?
8. Does the security force have its own communication system with direct lines between security and control areas?
9. Is there an auxiliary power supply for communication with each element of the security force?
10. Is sufficient equipment available to maintain continuous communications with each element of the security force?
11. What is the primary means of communication for the security force?
12. What is the alternate means?
13. Radio communications:
    a. Are proper radio procedures practiced?
    b. Is an effective routine code being used?

c. Is proper authentication required?
   d. Is the equipment properly maintained?
   e. Are frequencies dedicated solely for security force use? Do all guards know and recognize visual signals?
14. How often do the security officer and supervisory personnel review the firearms and ammunition requirements to ensure that they are adequate?
15. Is the security ammunition rotated?
16. Are guard posts reviewed and revalidated periodically?
17. Is overall supervisory responsibility for guard force operations vested with the security officer?
18. Is there a full-time guard supervisor for each shift?
19. Is a security center provided for guard force personnel supervisors?
20. Does the security center contain control equipment and instruments of IDS, warning, and guard communication systems?
21. Have direct communications with local municipal fire and police headquarters been established?
22. Are weapons stored in locked arms racks or containers when not in use?
23. Are periodic examinations conducted to ensure maintenance of guard training standards?
24. Is supervision of the guard force adequate?
25. Are general and special orders properly posted?
26. Do the guards know their general and special orders?
27. Are guard orders reviewed periodically to ensure applicability?
28. Are duties other than those related to security performed by security force personnel?
29. Does each guard on patrol duty at night carry a flashlight?
30. Do guards report or record their presence at certain key points:
    a. by means of a portable watch clock?
    b. by means of central watch clock stations?
    c. by means of telephone?
    d. by means of radio?
    e. not at all?
    f. by other means (identify)?
31. Are guard assignments, stationary locations, and patrol routes varied at frequent intervals to avoid establishing

routines or having the same individual performing the same duty assignment all the time?
32. In an emergency, how many other personnel are available to support the security force:
   a. at night?
   b. during the day?
   c. during weekends or holidays?
33. How many military personnel could be called during an emergency and would be able to arrive on-site:
   a. within one hour?
   b. within two hours?
34. Has a liaison been established with local, state, and/or federal law enforcement agencies such that early warning of threat situations will be provided or supported?
35. If a badge exchange system is used within any security area, does it provide for:
   a. comparison of badge, pass, and personnel?
   b. physical exchange of pass/badge at time of entrance or exit?
   c. accounting for each badge or pass?
36. Are all the badges that are maintained at the guard post entrance accounted for at the start and end of each shift?
37. What procedures are in place for the unaccounted badges?
38. When more temporary use badges are required by guard force personnel (due to a heavy influx of visitors, for example), are extra badges issued with receipts or simply handed to the guard force without accountability?

Utilities

1. What is the source of the electrical power for the facility?
2. Does the power enter the installation through underground cables or from overhead lines?
3. Where do the lines enter the installation?
4. What measures are taken to protect the power lines from tampering or sabotage?
5. Is sufficient power received to provide an ample reserve beyond the peak load demand?
6. Where are the primary transfer station(s) located?
7. How far are transfer stations from the facility perimeter?

8. How are they protected?
9. Where are the substations?
10. How are they protected?
11. Who is responsible for maintaining the power system?
12. How frequently is the power system checked and inspected?
13. If power is generated on the facility:
    a. Where is the powerhouse?
    b. How is the powerhouse protected from unauthorized access or sabotage?
    c. Where are the substations?
    d. How are they protected?
    e. When is the power system checked and inspected?
14. Are there any other measures taken to protect the power system from unauthorized use or sabotage?
15. What is the emergency power supply?
16. Can it be activated immediately?
17. Does the emergency power supply cover only critical areas of the facility?
    a. If so, what areas?
    b. What areas are excluded, if any?
18. What activities are supplied with emergency power?
19. Is the emergency power sufficient?
20. Who is responsible for emergency power?
21. Is the emergency power source located within a secure area? If so, indicate where and what protective measures are included.
22. Is there a main power switch for the entire facility?
    a. Who controls it?
    b. What measures are taken to protect it?
    c. Where is it located?
23. What is the source of water for the installation?
24. Is the source reasonably safe? Explain.
25. If water comes from a reservoir, tank, or the like, what is its capacity, water level, pressure, and condition?
26. How is it secured?
27. What is the water demand on the facility?
28. Is the source adequate compared to the demand?
29. Are pumping stations protected? If so, how?
    a. Where are they located?
    b. Are the pumps frequently inspected?

30. Are the water mains, control valves, and bypass valves protected?
31. How often is water tested for purity?
32. How is the test performed?
33. Is an emergency water source available? If so, describe its location, capacity, and potability.
34. What type of sewer system is used?
35. Does it appear to meet the needs of the facility?
36. Are there older sewer pipes that enter the facility and are no longer in use? If so, where are they located?
37. Can access to the facility be gained from these pipes?
38. If so, at what points do the pipes enter and exit the facility?
39. Have these pipes been blocked in some manner to preclude casual access?
40. What is the source of heat for the facility buildings?
41. Is central heating provided?
42. If so, where are the plants located and how are they protected?
43. Is the heating system periodically inspected?
44. How often?
45. Are maintenance personnel cleared if they have access to sensitive areas?
46. Is there an emergency system or series of emergency heating devices on the facility premises? Describe.
47. Where is fuel procured for the heating system?
48. Is it inspected for contamination? If so, what is the frequency of inspection?
49. How and where is it stored?
50. If rooms are individually heated, what measures are taken to prevent fires?

Vehicle Control

1. Are vehicles that are allowed routine access to the installation area registered with the security office?
2. Have written procedures been issued for the registration of all privately owned vehicles (POVs) authorized on the installation?
3. Is the vehicle identification pass easily discernible by security force?

4. Does the pass identify the specific agency/activity for whom the POV owner works?
5. Does the identification pass have an expiration date? Are the passes serialized?
6. If passes are not serialized, is there another method of determining whether or not a vehicle pass is valid for the installation?
7. Are separate passes used to determine the parking area? Where are they located on the POV?
8. What is the nature and scope of registration records maintained by the security office?
9. Is there a car or van pool system? Do registration records identify all vehicles belonging to each car or van pool?
10. Are there procedures to ensure that POV passes are invalidated when a POV is sold by an individual or when an individual departs?
11. Are trip tickets examined on official vehicles?
12. Are temporary passes issued to visitor vehicles?
13. Are visitor vehicles parked in an area specifically for visitors?
14. Are POVs allowed to park within security areas?
15. If not, are parking areas located away from sensitive points?
16. Are parking areas within any given security area located away from sensitive points?
17. Are parking areas such that individuals must pass through a pedestrian gate or other guard force control when they enter or leave work areas?
18. Are frequent or random searches made of POVs upon entering or leaving?
19. Are POVs authorized to park near security area fences?
20. Are the parking areas limited or specific, or is there a different designation for these types of vehicles? Are separate procedures in effect for freight vehicles?
21. Are procedures in effect for the issuance of parking permits for:
    a. emergency vehicles?
    b. VIP vehicles?
    c. courier vehicles?
    d. vendor vehicles?
    e. freight vehicles?

Appendix 2
Automated Information Systems Security Checklist

This checklist is designed to:

1. Provide an all-inclusive listing of specific areas of concern in the various facets of automated information systems (AIS) security;
2. Assist in determining areas of weakness that should be upgraded; and
3. Give managers, supervisors, and operating personnel a better grasp of AIS security as it concerns their job, associated functions, and the relationship to the working environment.

The checklist has been separated into two sections; the first may apply to any AIS area, while the second is designed specifically for government—in this case the Department of Defense (DOD). The second section has been included because many government activities have AIS, with the DOD being the major user. Because of national security concerns in the protection of data, DOD has developed a specific checklist.

Do not limit the inspection process to an individual section, but use the best of both sections to develop a local site checklist.

It is anticipated that future requirements, systems hardware/modifications, and new viewpoints toward increased security may require this listing to be updated. Therefore, every individual must be aware of potential security weaknesses and identify them.

Acronyms used in this checklist are:

ADP—Automatic Data Processing

ADPE—Automatic Data Processing Equipment

AIS—Automated Information Systems

AIS SSO—AIS System Security Office/Officer

CI—Counterintelligence

ISP—Information Security Program (DOD Regulation 5200.1-R)

SSO—System Security Office/Officer

ST&E—Security Testing and Evaluation

TSCM—Technical Security Countermeasures

SECTION 1—ALL AUTOMATED INFORMATION SYSTEMS FACILITIES

System Security Procedures—General

1. Are the procedures designed to ensure that system security is well documented?
2. Are the procedures reviewed and updated regularly?
3. Is there an effective procedure for ensuring that batch job and data inputs have been prepared and submitted by authorized system users?
4. Are these inputs logged as to date and time of submission, job number, and person submitting?
5. Is there an effective procedure for ensuring that all computer system outputs are delivered to the properly authorized system user?
6. Are all such output materials logged out with regard to date and time, job number, and recipient?
7. Is there an effective procedure to ensure that all system outputs are properly marked to reflect security classification and

any applicable access categories and that such markings are in accordance with appropriate regulations?
8. Is there a procedure for controlling classified output from the time it is created to the time it is destroyed?
9. Is a complete system operations log maintained, showing all system activities, problems, the date and time they occurred, and the name of the individual making each entry?

System Certification and Testing

1. Has the approving authority developed a complete set of parameters for System Testing and Evaluation (ST&E), certification, and retest and recertification of AIS systems?
2. Does the ST&E plan provide for a detailed examination of all security safeguards—such as hardware; software; and procedural, administrative, operational, counterintelligence, and physical security—to estimate the probability of inadvertent disclosure of classified information?
3. Are ST&E conducted by outside personnel composed of a mix of computer and selected security specialists?
4. Are there procedures in effect to request inspections, technical surveillance countermeasures, physical security, and emanations security inspections prior to the start of classified operations and periodically thereafter?
5. Are there tests to determine that security safeguards incorporated into hardware and software are operative, function as intended, and constitute acceptable controls?
6. Do certification tests verify that the security features designed into the equipment are in fact present, work effectively, and constitute adequate security safeguards? Do these tests also address operational procedures, administrative structure, environmental controls, and physical safeguards?
7. Are recertification tests conducted periodically, after a system malfunctions, and after scheduled or unscheduled hardware or software maintenance or modification?
8. Are all security parameters inserted into the system by the SSO?
9. Are tests conducted to ensure that information is passed to or accepted from any portion of a system only at the level authorized for that component?

10. Are postcertification changes authorized only by the SSO and conducted by systems software or hardware personnel officially designated to perform such functions?
11. Are explicit reports of all such changes made to the certifying authority?
12. Are the following records maintained to assist in audit trails:
    a. User log-ons and log-offs including terminal and user ID?
    b. Maintenance actions with name of maintenance personnel, nature of maintenance, and files accessed?
    c. Operator-initiated functions with name of operator and function?
    d. All attempts by a user to access files for which he or she does not have proper authorization?
    e. All program abort incidents?
    f. All special uses, generation of passwords, changes of classification, or modification of security parameters?
    g. Do AIS SSO logs contain sufficient detail to reconstruct all unsuccessful attempts to penetrate the system?
    h. Do operator logs support the AIS SSO log on details of system operation?

Personnel Security

1. Are all personnel properly cleared for information within the system?
2. Are proper clearance records maintained?
3. Are clearances updated when required?
4. Are access rosters maintained in restricted areas?
5. Have the AIS SSO and operations supervisors been briefed on the "personal" in personnel security?
6. Have supervisors and the AIS SSO been alerted to watch for character deficiencies or unexplained affluence?
7. Is maximum use made of the principle of need-to-know?
8. What steps are taken to assure that the computer user possesses the proper clearances and special accesses for the information and data at his or her disposal through the AIS system(s)?
9. Are all computer systems personnel required to take their earned annual leave?

10. Are computer systems personnel required or allowed to work more than eight hours per day, five days per week for excessively long periods of time?
11. Are people cross-trained to cover all functions?
12. Is there any procedure or effort aimed at detecting and dealing with an employee's bad attitude or personal problems that affect job performance?
13. Do supervisors advise security personnel of possibly disgruntled employees?
14. Is there an established procedure for resolving employee grievances?
15. Is so, does it generally produce results deemed just and fair by the employee?
16. What is the attitude of the employees toward their agency, department, branch, supervisor, duties, and so forth?
17. Are the employees generally satisfied with their compensation, working hours, shift assignments, and the like?
18. Do supervisors know their employees well enough to detect changes in their living habits?
19. Do personnel policies allow for containment or job reassignment to a nonsensitive position for employees who may constitute a threat to the AIS installation?
20. Is there a continuing personnel education program in computer security matters?
21. Have users been educated to the point that they are convinced of the need for existing computer security measures and they accept the burden imposed by these measures without complaint?

Document and Information Security

1. Is the security control system sufficient to ensure that no one can obtain access to classified information within the system without proper authorization?
2. Have approved security containers or the equivalent been provided for all demountable media?
3. Has the AIS equipment area been constructed as a vault and approved for open storage of classified information?

4. If the AIS equipment area has not been approved for open storage, are all nondemountable media items containing classified material properly stored or placed under continuous surveillance?

Marking and Handling

1. Are hard-copy input/output products marked with applicable classification and downgrading markings?
2. Are magnetic media such as memories, tapes, discs, and drums properly marked and handled?
3. Are classified machine components, including program circuit boards, properly marked to indicate classification?
4. Are classification labels utilized?

Transmission and Dissemination

1. Are detailed procedures published on the transmission of all types of AIS media?
2. Are proper packing and receipting procedures for transmitting AIS media being followed?

Need-to-Know

1. Does the security system make maximum use of compartmentation and restrict the information to only that required?

Security Control Personnel

1. Has an AIS Systems Security Officer been appointed in writing?
2. Is the AIS SSO a computer operations specialist who is trained in security functions?
3. Does the AIS SSO have direct access to the unit or activity commander?
4. Is the AIS SSO thoroughly trained in his or her duties?
5. Do supervisory personnel in programming, machine operations, administration, remote sites, and the library assist the AIS SSO in security control procedures?

6. Are all of these personnel indoctrinated on the importance of their security control functions?
7. Is there a terminal area security officer appointed in writing for each remote terminal site?

Document Accountability Systems

1. Has a detailed procedure for accounting for classified information been prepared and implemented?
2. Does the system accurately account for classified information during its entire life cycle from generation to destruction?
3. Does the system account for magnetic storage media that have been erased or degaussed?
4. Have separate and positive control measures been established for certified copies of the supervisory software?

Destruction or Declassification

1. Have destruction and declassification procedures for hard-copy media been prepared and implemented?
2. Are they effective?
3. Is all material completely destroyed or accounted for?
4. Are downgraded copies properly marked?
5. Are degaussing procedures being followed for magnetic storage media?
6. Are stringent verification procedures used on all erased or degaussed material?

Technical Surveillance Countermeasures

1. Were TSCM technicians consulted for preconstruction technical assistance during the planning phase?
2. Were TSCM standards included in the original construction plans?
3. Was a TSCM survey conducted after the site became operational?
4. Was a TSCM survey conducted during the past three years?

5. If technical hazards were indicated during the technical survey, have they been corrected?
6. Have personnel been cautioned to discuss classified information only in properly cleared areas?
7. Have all personnel been instructed to leave any suspected clandestine listening devices undisturbed?
8. Have all personnel been instructed not to discuss their suspicion that an object might be a technical surveillance device within the range of the device?
9. Do all personnel know whom to contact if they suspect a technical surveillance device?

Countersabotage

1. Are physical access control measures designed to thwart external sabotage attempts in effect?
2. Do these access control measures extend to all supporting equipment?
3. Are supervisory personnel alert to the detection of the disgruntled employee who might cause internal sabotage?

Systems Design

1. Has security been integrated into systems design planning?
2. Has redundancy of equipment and operations been considered during systems design planning?
3. Has systems design attempted to make the system flexible?
4. Has it attempted to make the system responsible?
5. Auditable?
6. Reliable?
7. Adaptable?
8. Dependable?
9. Has systems design considered configuration integrity?
10. Has systems design considered that the equipment must self-test, violate its own safeguards deliberately, attempt illegal operations, and monitor user actions?

Hardware Security

1. Are hardware security features designed to protect the supervisory software, isolate users, and detect unanticipated conditions?
2. Are there self-testing features, to include deliberate attempts to bypass the security controls?
3. Do computer operator personnel report failures to the AIS SSO?
4. Is entrance into the privilege mode hardware-controlled?
5. Are positive hardware measures employed to separate privilege and user modes?
6. Do hardware controls return the system to the privilege mode if unauthorized attempts are made to enter the supervisor?
7. Do hardware safeguards provide for predicted responses to unanticipated conditions?

Software Security

1. Are the software functions of utility routines, language processors, and the program library effective?
2. Are they repeatedly tested for proper function?
3. Are terminal, user, and file authenticators or passwords used?
4. Are they effective?
5. Are they properly safeguarded from unauthorized personnel?
6. Are supervisor functions separated into self-contained modules with explicit instructions governing communication between modules?
7. Do software controls assure that no program-accessible residue remains in either primary or secondary storage?
8. Are software security functions continually self-testing and reporting?
9. Are software controls such that the user must absolutely identify him- or herself and authenticate his or her identity?
10. Do the software controls check an individual's degree of access and the degree of access of the terminal he or she is using?
11. Does such separately identifiable information contain an access authorization requirement?
12. Does the software prevent access by unauthorized users?

13. Does the software provide for a system of holding and return to the user if the user makes a legitimate operational error?
14. Does the software provide for assistance in determination of file classification?
15. Are software traps incorporated to detect any input or output identified by a security flag that exceeds that authorized for the user, the user's terminal, or any file specified in his or her job?
16. Does software automatically provide for the change of security flag when data from a file with a different security flag is included?

Tape Library

1. Magnetic tapes and disks. Does the tape and disk accountability procedure cover:
   a. frequency of use?
   b. frequency of cleaning?
   c. authorized user?
2. Is there an effective procedure for controlling access to removable media such as magnetic tapes and disk packs?
3. In each instance, do operations personnel ensure that the user job requesting access is in fact authorized access?
4. Is there an effective procedure to ensure that all previously used magnetic storage media have been effectively erased before they are reassigned to a new user?
5. Are the proper declassification procedures followed when computer storage materials are to be released and possibly subjected to laboratory analysis by personnel not authorized access to the data they once contained?
6. Are magnetic tapes and disks filed in an orderly manner?
7. Do you check and clean your disk packs or have it done by contract?
8. Are tapes kept in their containers except when in use?
9. Does the facility use a pass/badge exchange system?
   a. If so, do both badges bear identical photographs?
   b. Are both printed from the same negative, taken at the same time?
10. Does the guard compare the pass/badge to the bearer and also to the pass/badge being exchanged?

11. Must the pass/badge be exchanged each and every time an individual enters and departs the facility during the same day?
12. Is there an accounting for each badge/pass?
13. Is an inventory of badges and passes conducted by security personnel at the beginning and end of each tour of duty?
14. What is the security of badges not in use?
15. Is there a location check of personnel who have not checked out of the facility/area at the close of daily operations?
16. How are visitors to the facility identified?
17. Through what or how many entrances are visitors permitted to enter? How is this enforced?
18. Are visitors required to enter and exit through the same entrance?
19. Are signs containing instructions for visitors posted at facility entrances?
20. Is an escort system utilized for visitors? If so, is there an escort pool?
21. How many personnel are assigned to the escort pool?
22. Is this adequate for the number of visitors to the facility?
23. How are escorts assigned to the visitors?
24. Are escort personnel permanently assigned?
25. Are escorts aware of the areas into which visitors are not allowed?
26. How are escort personnel identified?
27. Does the escort remain with the visitor during the entire period of the visit?
28. If not, what other control measures are employed?
29. What action is taken if the escort and visitor are separated?
30. Are escorts provided by the offices of the individuals to be visited?
31. If so, how are visited personnel notified of visitors?
32. How does the guard determine that the person desiring entry is an authorized visitor?
33. Is a record maintained to reflect both the visitor and the escort?
34. How long is the record maintained?
35. Does the escort accompany the visitor to the entrance upon completion of the visit? If not, what procedures are followed?
36. Are special visitor badges used for VIPs?
37. Are special or distinctive visitor badges used for vendors, utility, or other servicing personnel?

38. Are VIP visitor controls the same as for any other visitor? If not, what procedures are used?
39. Do permanent employee badges have an expiration date? If not, why not?
40. Are permanent employee badge pictures checked to determine if there has been a major change in physical appearance?
41. If so, is a new badge required? If not, why not?
42. What deterrents to alteration or reproduction have been incorporated into the facility badge?
43. Is any coding system incorporated into the badge to designate areas or gates of access, shifts, and so on?
44. Are any electronic or automatic identification systems utilized?
45. If so, are badges spot-checked to ensure that a badge belongs to an authorized bearer?
46. Have measures been taken to safeguard the terminal identification of remote terminals?
47. Are all personnel entering any terminal area verified with regard to security clearance and need-to-know?
48. Are uncleared visitors continually escorted while in the terminal area?
49. Are terminal personnel indoctrinated to report suspicious behavior or security infractions on an annual basis?
50. Are all maintenance personnel cleared for the highest level of information contained in the system or placed under an appropriate escort who is technically qualified?
51. Is the identity of all maintenance personnel positively verified?
52. In systems that process highly sensitive data, is maintenance carried out under the two-man rule?
53. Are logs maintained on all maintenance actions?
54. If maintenance actions require modification of security safeguards, are procedures in effect to require reinspection and recertification?
55. Is on-line debugging of supervisor software prohibited except when all on-line storage devices containing classified files not needed for maintenance are disconnected from the line?
56. Does the continuity of operations plan provide for alternate operational sites, duplicate copies of all critical data, programs, and security procedures?

57. Have these plans, equipment, and data been periodically exercised to maintain operational readiness?
58. Do management controls specify procedures for the maintenance of environmental controls of the computer room area?
59. Are procedures in effect to solve such problems as brownouts, blackouts, and transients in the power system?
60. Is auxiliary power available?
61. Are fire control procedures and equipment available?
62. Have periodic drills been conducted?
63. Do all personnel know their functions during such drills?
64. Have measures been taken to protect the AIS equipment from water damage in the event of fire or flood?
65. Are visitors to the AIS area prohibited except on absolutely essential business?
66. Does the AIS area have a viewing area for VIPs?
67. Does this area preclude viewing classified material or operations when requested?
68. Is there a backup chilled water system?

Assignment of Security Duties

1. Is the responsibility for carrying out security procedures specifically assigned to appropriate supervisory personnel?
2. Are the personnel who are principally involved with ensuring the security of the computer facility in the same branch or department as the computer system operations personnel or the functional users of the computer system?
3. Are security duties relegated to operations or functional user personnel during nonbusiness hours?
4. Are computer operations personnel normally made sufficiently aware of the computer security problem and its importance?
5. Are computer operations personnel restricted from making program modifications and systems patches?
6. Is there a security training program specifically tailored to operations personnel? Are operations personnel able to deal with unanticipated problems having security implications?
7. Who is responsible for system patch validation and what steps are taken?

8. Is there an effective program to make all personnel involved in the computer operation aware of their duties and responsibilities regarding security?
9. Are there procedures to ensure that all personnel receive the AIS training they need to carry out their jobs properly?

Physical Security

1. Is the computer facility adequately secured during the hours it is not in use?
2. Are guard posts properly manned so that entrance to the data processing facility will not be unguarded at any time?
3. Is access to floors above and below the computer facility and to areas adjacent to the facility restricted and controlled?
4. What controls exist over the issuance of keys, combinations, and the like? Are the locks and/or combinations changed upon termination of employment of personnel who had access to them?
5. When computer facility doors are operated by push-button combinations, are the controls sufficient to preclude the acquisition and use of combinations by unauthorized persons?
6. When push-button combination locks are used, are the combinations changed periodically?
7. When this type of lock is used, will an alarm be given if an improper combination is entered?
8. Are the physical safeguards of the system in accordance with the requirements for the highest level of data being processed?
9. When more than one computer system occupies the same physical area, are the proper safeguards taken to ensure that they are electrically and informationally isolated from each other?
10. Are visitor logs maintained at the entrance used by visitors?
11. Does the visitor log show all deliveries by messengers, as well as what was delivered and to whom?
12. Are visitor clearance levels adequately checked?
13. Are visitors issued badges that clearly indicate they are visitors and are cleared to a specified level?

14. When visitors with a given clearance level are present, are proper precautions taken to shield from view all classified materials of a higher level than the visitor's clearance level?
15. Is access to the computer room itself denied to all but authorized management and operating personnel?
16. Do all personnel with unescorted access to the system possess a clearance/special access authorization equal to or higher than the highest classification and all categories being processed?
17. Are all computer operators and system programming personnel cleared for the highest level and most restrictive category of classified information in the system?
18. Are the computer facility access controls enforced with equal vigor during nonbusiness as well as business hours?
19. Is there a badge system specifically for control of access to the computer facility?
20. If so, does the badge contain the employee's photograph and signature and indicate his or her level of clearance?
21. Are the badges coded to indicate access privileges for restricted areas?
22. Are visitors' packages inspected before being permitted into the computer facility?
23. Are visitors properly escorted when their clearance level is lower than the system's highest level?
24. Are the same stringent access controls in force for service personnel (cleaning crew, maintenance men, telephone repairmen, messengers, and so forth) as for regular operating personnel?
25. Are file areas segregated so that only specific individuals have access?
26. Are keys, cypher locks, and other security devices used to control access?
27. Are keys and locks changed at regular intervals or after termination of an employee?
28. Are employees who are dismissed from the computer environment removed immediately, their admission badges retrieved, and the necessary guard personnel notified?
29. Are keys controlled? Signed for? Turned in when an employee departs?

30. Are personnel trained to challenge improperly identified or badged visitors?
31. Is there a visitor control procedure?
32. Are escort procedures established for controlling visitors?
33. Are all potential escorts properly briefed on their responsibilities?
34. Can you prevent a single individual from gaining access during off-shift hours without the knowledge of a security guard or other employee?
35. Is the central computer facility manned by at least two appropriately cleared personnel at all times?
36. Is advertising the location of the computer discouraged?
37. Is there standby battery power to operate electrically controlled doors during power failures?
38. Are the physical barriers such as walls, doors, alarms, and guard posts sufficient to deter a determined individual from attempting to penetrate the facility?
39. Are these barriers constructed in depth so that no single barrier is used to protect any component of the system?
40. Has the area been designated a "Restricted Area"?
41. Have the locks, intrusion detection systems, and other access controls been subjected to attempted penetrations by security personnel to determine strengths and weaknesses of the system?
42. Are remote terminals protected to the highest level of classified information that they process?
43. Do access controls prevent the introduction of magnets or electronic equipment into the computer room area?
44. Are the access controls continuous to prevent tampering or removal of items during down time?

Fire Exposure and Risk

1. Is the computer housed in a building that is constructed of fire-resistant and noncombustible materials?
2. Is the computer room separated from adjacent areas by noncombustible fire-resistant partitions, walls, and doors?

3. Are paper and other combustible supplies stored outside the computer area?
4. Are file tapes and disks stored outside the computer area?
5. Are flammable or otherwise dangerous activities prohibited from adjacent areas or areas above or below the computer room?
6. Is raised flooring made of noncombustible material?
7. Are operators trained periodically in firefighting techniques and assigned individual responsibilities in case of fire?
8. Are curtains, rugs, furniture, and drapes noncombustible?
9. Is smoking restricted in the computer area?
10. Is the computer area protected by (not in any order of preference):
    a. Automatic carbon dioxide?
    b. Personnel trained in the use of gas masks and other personal safety measures?
    c. Halogenated agents?
    d. Water?
    e. Wet pipe (releases water at a set temperature)?
    f. Preaction (may sound alarm and delay water release)?
11. Are portable fire extinguishers spread strategically around the area with location markers clearly visible over computer equipment?
12. Are emergency power shutdown controls easily accessible at points of exit?
13. Does emergency power shutdown include air conditioning?
14. Are smoke detectors installed:
    a. in the ceiling?
    b. under the raised floor?
    c. in air return ducts?
15. Does smoke detection equipment shut down the air conditioning equipment/system?
16. Are smoke detectors properly engineered to function in the computer room?
17. Is the smoke detection system tested on a regularly scheduled basis?
18. Are fire drills held regularly?
19. Is a cleaned floor maintained under the raised floor?
20. Is there an adequate supply of firefighting water available?

21. Is there battery-powered emergency lighting throughout the computer area?
22. Are there enough fire alarm pull boxes within the computer area and throughout the facility?
23. Does the alarm sound:
    a. locally?
    b. at security guard station?
    c. at central station?
    d. at fire or police headquarters?
24. How does the local firefighting force rate according to the American Insurance Association's Standard Fire Defense Rating Schedule?
25. Is there round-the-clock security coverage during non-working hours?
26. Are the flammable materials used in computer maintenance, such as cleaning fluids, kept in small quantities and in approved containers?
27. Can emergency crews gain access to the facility without delay?
28. If access is via an electrically controlled system, can it be operated by standby battery power or is there a hand-operated bypass mechanism?

Water Damage Exposure

1. Are computers excluded from areas below grade?
2. Is there an effort to eliminate any overhead steam or water pipes except for sprinklers?
3. Is there adequate drainage under the raised floor?
4. Are drains installed on the floor above to divert water accumulations away from all hardware?
5. Is there adequate drainage to prevent water overflow from adjacent areas?
6. Are all electrical junction boxes under the raised flooring held off the slab to prevent water overflow from adjacent areas?
7. Are exterior windows (if any) and doors watertight?
8. Is there protection against accumulated rainwater or leaks in rooftop cooling towers?

Air Conditioning

1. Is the system used exclusively for the computer area?
2. Are duct linings noncombustible?
3. Are filters noncombustible?
4. Is the compressor remote from the computer room?
5. Is the cooling tower fire protected?
6. Is there backup air conditioning capability?
7. Are air intakes:
   a. covered with protective screening?
   b. located well above street level?
   c. located to prevent intake of pollutants or other debris?
8. Are air temperature and humidity recorded in the computer environment?

Electrical Power

1. Is uninterrupted power required because of the nature of the facility business operations?
2. If the system requires motor generators, is there a backup?
3. Has the local power supply been checked for reliability?
4. Has the power source been monitored with recorders to identify electrical transients?
5. In the event of power failure, is there emergency lighting for removal of personnel?
6. Is backup power tested at regular intervals?
7. Are there lightning arrestor(s)?
8. Is there emergency power off at all exits and within the computer center?
9. Are emergency power offs protected from accidental activation?

Natural Disaster Exposure

1. Is the building structurally sound:
   a. to resist flood damage?
   b. to resist windstorms and hurricanes?
   c. to resist earthquakes?

2. Are the building and equipment properly grounded for lightning protection?
3. Is the building on a solid foundation?
4. Is the building remote from any earthquake faults?

SECTION 2—DEPARTMENT OF DEFENSE FACILITIES

The Department of Defense, similar to other government agencies, prepares an initial inspection checklist for use by the various military departments. It is included here. Several notes for consideration, though, are in order:

- This is a "baseline" checklist in that it does not go into detail, but covers the minimum level of inspection required under the program.
- Other government agencies may use a similar checklist but, except for the Department of Energy, they tend to duplicate each other, mostly relying on the DOD listing.
- To use this listing, look at the other more specialized listings and make appropriate changes so that the final checklist does more than meet the minimum standards. Using other listings ensures completeness in developing and implementing a site-specific PSI checklist.
- Allow several blank lines between questions for written comments, as appropriate, that arise during the inspection.

Program Management

1. Does the activity have DOD 5200.1-R and any component/activity supplementing instructions to provide for the internal administration of the Information Security Program?
2. Has a senior information security authority been designated in writing by the head of the component to be responsible for implementation of the Information Security Program (ISP) within the component?
3. If the activity is involved with major projects or programs, does it have a program security classification guide? Does

it hold a copy of DOD 5200.1-I, "DOD Index of Security Classification Guides," and DOD 5200.1-H, "DOD Handbook for Writing Security Classification Guides"?
4. Has an official been appointed in writing by the head of an activity to serve as security manager and is he or she responsible for the administration of an effective ISP in that activity, to include sampling of classified documents?
5. Does the security manager have the necessary training to perform the job?
6. Does the security manager have direct and ready access to the appointing official?
7. Do component Senior Information Security Authorities ensure that activities are monitored and inspected, with or without prior announcement, and reports completed on the status of administration of their ISPs?
8. Have a Top Secret Control Officer and alternate been designated in writing where there is likelihood of processing Top Secret information, and are they performing their assigned responsibilities?
9. Is the activity collecting and reporting data on the SF 311, "Agency Information Security Program Data," to satisfy the report requirements of the Information Security Oversight Office (ISOO)?
10. Are any significant requirements levied directly on a component by ISOO brought to the attention of the Director of Security Plans and Programs, Office of the Deputy Under Secretary of Defense (Security Policy)?
11. Has each component head established and maintained an ISP that is designed to ensure compliance with the provisions of DOD 5200.1-R throughout the component?
12. Are the differences between the terms "Limited Dissemination" and "Special Access Program" clearly understood?

Classification Policies and Principles

1. Is (original) security classification applied only to protect the national security and only as long as required by national security considerations?

2. If there is a reasonable doubt about the need to classify information, is it safeguarded as appropriate pending a determination by an original classification authority?
3. Are other terms such as "sensitive" avoided in conjunction with the authorized classification designations to identify classified information?
4. Are delegations of original classification authority limited to the minimum number required for efficient administration and to those officials whose duties involve the origination and evaluation of information warranting classification at the level stated in the delegation?
5. Are requests for the delegation of original classification authority made only when there is a demonstrated and continuing need to exercise such authority?
6. Are original classification authorities indoctrinated in the fundamentals of security classification, limitations on their authority to classify information, and their responsibilities as such prior to the exercise of this authority?
7. Are listings of original classification authorities reviewed at least annually to ensure that the officials so listed have demonstrated a continuing need to exercise original classification authority?
8. Do persons who have a derivative classification responsibility verify the information's current level of classification as far as practicable before applying the markings?
9. Have any other officials, in addition to original classification authorities, been designated to exercise declassification and downgrading authority over classified information in their functional areas of interest?
10. Do original classification authorities normally apply the six-step process for their original classification determinations?
11. Do original classification authorities abide by the limitations on classification?
12. Are documents classified on the basis of the information they contain or reveal?
13. Is the need for continued classification reevaluated when classified information has been lost or possibly compromised, including appearance in the public domain?

14. In unusual circumstances, is classification by compilation of unclassified items of information fully supported by a written explanation that is required to be provided with the material so classified?
15. Is information extracted from a classified source derivatively classified or not classified in accordance with the classification markings shown in the source?
16. When it can be determined, do original classification authorities set dates or events for declassification at the time the information is classified originally?
17. When extending the duration of classification, which had been specified at the time of original classification, do officials with the requisite original classification authority take this action only if all known holders of the information can be notified of such action before the date or event previously set for declassification?
18. Is any decision to continue classification of information designated for automatic declassification under previous executive orders, or other than on a document-by-document basis, reported to the Deputy Under Secretary of Defense for Security Policy [DUSD(SP)]?

Classification Guides

1. Is advance classification planning considered essential and does the official charged with developing any plan, program, or project—in which classification is a factor—include under an identifiable title or heading, classification guidance covering the information involved?
2. Is a classification guide issued for each classified system, program, plan, or project as soon as practicable before the initial funding or implementation of the system, program, plan, or project?
3. Do successive operating echelons prescribe more detailed supplemental guides that are considered essential to assure accurate and consistent classification?
4. Do guides identify the information elements to be protected?

5. Are particular levels of classification and declassification instructions for each element or category of information specified in terms of a period of time, the occurrence of an event, or a notation that the information shall not be declassified automatically without approval of the originating agency?
6. Are any special public release procedures and foreign disclosure considerations specified in the guide?
7. Is each classification guide approved personally and in writing by an official who has program or supervisory responsibility over the information or is the senior agency official designated in accordance with DOD 5200.1-R and authorized to classify information originally at the highest level of classification prescribed in the guide?
8. Is a program security classification guide developed for each system and equipment development program that involves research, development, test, and evaluation (RDT&E) of classified technical information?
9. Whenever possible, do classification guides specifically cover each phase of transition—that is, RDT&E, procurement, production, service use, and obsolescence—with changes in assigned classifications to reflect the changing sensitivity of the information involved?
10. Do originators review classification guides for currency and accuracy at least once every two years, and if no changes are made, is the record copy annotated to show the date of the review?
11. Are classification guides, including those in current use, updated to meet the ISP requirements?
12. Is a copy of each approved classification guide and changes thereto—other than those covering sensitive compartmented information (SCI) or a Special Access Program and those disclosing information that requires special access—sent to the Director of Freedom of Information and Security Review, Office of the Assistant Secretary of Defense (Public Affairs), and to the Director of Security Plans and Programs, Office of the Deputy Under Secretary of Defense for Security Policy [ODUSD(SP)]?
13. Unless each approved guide is classified Top Secret or covers SCI or is determined by the approval authority of the guide to

be too sensitive for automatic secondary distribution to DOD components (such as a Special Access Program guide revealing the nature of the program), does the originator send two copies and changes thereto to the Administrator, Defense Technical Information Center (DTIC), Defense Logistics Agency?
14. Does each guide provided to DTIC bear a new distribution statement from DOD Directive 5230.24, "Distribution Statements on Technical Documents?"
15. Do originators of each guide execute a DD Form 2024, "DOD Security Classification Guide Data Elements," when the guide is approved, changed, revised, reissued, or canceled, and when its biennial review is accomplished?
16. Is the original of each executed DD Form 2024 forwarded to the Director, Security Plans and Programs, ODUSD(SP), for listing in DOD 5200.1-I, "DOD Index of Security Classification Guides" that is maintained by ODUSD(SP)?
17. Are classification guides, which are not listed in DOD 5200.1-I and do not require an executed DD Form 2024 because of classification considerations, reported separately to the Director of Security Plans and Programs, ODUSD(SP)?

Classification Challenges/Resolution of Conflicts

1. If holders of classified information have substantial reason to believe that information is classified improperly or unnecessarily, are they required to communicate that belief to their security manager or the classifier of the information to bring about any necessary correction?
2. Have procedures been established whereby holders of classified information may challenge a decision of a classifier that will not result in or serve as a basis for adverse personnel action?
3. Are classification disagreements resolved promptly and in accordance with standard procedures?

Classification in Industrial Operations

1. Is the DD Form 254, "Contract Security Classification Specification," used by DOD contracting activities to convey contractual security classification guidance to industrial management?

2. Are DD Forms 254 changed by originators to reflect changes in classification guidance and reviewed for currency and accuracy at least once every two years?
3. Do changes in guidance for the DD Form 254 conform with DOD 5200.1-R, appropriate industrial security regulations and procedures, and are changes provided to all holders of the DOD Form 254 as soon as possible?
4. When no changes are made as a result of the biennial review of the DD Form 254, does the originator notify all holders in writing?

Declassification and Downgrading

1. Are decisions concerning declassification based on the loss of sensitivity of classified information with the passage of time or on the occurrence of an event that permits declassification?
2. Are proper procedures followed with respect to requests for mandatory reviews of DOD classified information?
3. Whenever practicable, are classified documents reviewed for downgrading or declassification before they are forwarded to a records center for storage or to the National Archives and Records Administration for permanent preservation?
4. When declassification action is taken earlier than originally scheduled, or the duration of classification is extended, does the authority for making such changes ensure prompt notification to all holders for whom the information was originally transmitted?

Marking

1. Are originally classified documents marked properly?
2. Are derivatively classified documents marked properly?
3. Does the activity hold copies of DOD 5200.1-PH, "A Guide to Marking Classified Documents"?
4. Is the classification authority properly identified on classified documents?
5. Is the original classification authority identified on the "classified by" line if all the document's information is classified as an act of original classification?

6. Is "multiple sources" listed on the "classified by" line if the classification is derived from more than one original classification authority, or an original classification authority and another source, or from more than one source document, classification guide, or combination thereof?
7. When "multiple sources" are listed on the "classified by" line, is a listing of these sources fully identified and maintained with the file or record copy of the document?
8. Are documents properly marked with the overall classification, including page marking, and, except for blank pages, are interior pages marked according to their individual content including "unclassified" when no classified information is contained on such a page?
9. Are major components, namely, annexes and appendices, of complex documents properly marked?
10. Are document portions marked?
11. Are illustrations, photographs, figures, graphs, drawings, charts, and similar portions of classified documents, as well as captions of such portions, properly marked?
12. Are documents containing compilations of unclassified information warranting classification and compilations of unclassified portions within documents properly marked with the overall classification that would include an explanation of the basis of the assigned classification?
13. If subjects or titles of classified documents cannot be made unclassified, are those that are classified properly marked?
14. Are classified document cover sheets affixed to classified documents that have been removed from secure storage?
15. Are transmittal documents properly marked?
16. Are electronically transmitted messages marked properly?
17. Are markings properly applied on special categories of material?
18. Are transparencies and slides marked properly?
19. Are motion picture films and video tapes marked properly?
20. Are interior pages of fan-folded computer printouts marked properly?
21. Whenever classified information is downgraded or declassified earlier than originally scheduled, or upgraded, is the material properly remarked?

22. Are derivative declassification dates properly applied?
23. Is each classified document marked on its face with one or more of the standard markings?
24. When individual documents or materials are permanently withdrawn from storage units, are they promptly and properly remarked?
25. Is classified information that an originator has determined is subject to special dissemination or reproduction limitations marked with an appropriate statement or statements, as appropriate?
26. Have additional warning notices been properly applied?
27. Are originators using dissemination and reproduction notices to support unapproved Special Access Programs?
28. Is the remarking of material prepared under previous executive orders in accordance with proper current procedures?
29. Have all action officers or other personnel who originally or derivatively prepare documents containing classified information been trained in the original and derivative classification processes and in downgrading and declassification procedures?

Storage and Storage Equipment

1. Is information and material afforded the level of protection against unauthorized disclosure commensurate with the level of classification assigned, and is this information stored only under conditions adequate to prevent unauthorized persons from gaining access?
2. Is classified information that is not under the personal control and observation of an authorized person guarded or stored in a locked security container as prescribed for the various levels of classification?
3. Is a preliminary survey conducted prior to the procurement of new storage equipment, and is new storage equipment procured from those items listed on the GSA Federal Supply Schedule?
4. Do vaults or containers bear externally an assigned number or symbol in lieu of an external mark referring to the level of classified information authorized to be stored therein?

5. Are combinations to security containers changed only under certain conditions and at least annually?
6. Are combinations classified at the highest category of the classified information authorized to be stored therein?
7. Is a record, namely, Standard Form 700, "Security Container Information," maintained for each vault, secure room, or container used for storing classified information, showing location of the container, the names, home addresses, and home telephone numbers of the individuals having knowledge of the combination?
8. Is access to combinations limited to only those individuals who are authorized access to the classified information stored therein?
9. Are electrically activated locks prohibited for the storage of classified information?
10. Are repairs to damaged security containers performed according to proper procedures?

Custodial Responsibilities

1. Are custodians responsible for providing protection and accountability for classified information?
2. Are procedures established to prohibit the removal of classified material from an activity to do homework or for other reasons, unless approved by the head of a DOD component or single designee at the headquarters and major command levels and then only when a GSA-approved security container is furnished to safeguard the material?
3. When classified documents are removed from storage, are they kept under constant surveillance and face down or covered when not in use with cover sheets, namely, Standard Forms 703, 704, and 705 for, respectively, Top Secret, Secret, and Confidential documents?
4. Are preliminary drafts, typewriter ribbons, sheets, stencils, and the like, protected according to their content and destroyed after they have served their purpose?
5. Has a system of security checks at the close of each working day been established to ensure that the area is secure, and are

Standard Forms 701, "Activity Security Checklist," and 702, "Security Container Check Sheet," used as part of this system?
6. Have emergency destruction and evacuation plans been developed and tested?
7. Is there a prohibition against discussing classified information over the telephone?
8. Are security requirements and procedures governing disclosure of classified information at conferences, symposia, conventions, and similar meetings, and the requirements and procedures governing the sponsorship and attendance of U.S. and foreign personnel at such meetings followed?
9. Except for classified information that has been officially released to the custody of a foreign country, is the retention of U.S. classified material in foreign countries authorized only when that material is necessary to satisfy specific U.S. government requirements, and the material is stored under U.S. government control under certain conditions?
10. Has an activity entry and exit program been established to deter unauthorized removal (and introduction) of classified material?

Compromise of Classified Information

1. When the loss or possible compromise of classified information has occurred, is appropriate investigative and other action taken to identify the source(s) and reason(s) for the compromise and remedial action(s) taken to ensure further compromise(s) do not recur?
2. Is the originator notified and requested to conduct a review and reevaluation of the information subjected to compromise?
3. Has each DOD component established a system of controls and internal procedures to ensure that damage assessments are conducted when required?
4. Are espionage cases and deliberate unauthorized disclosures of classified information to the public reported in accordance with DOD Instruction 5240.4, "Reporting of Counterintelligence and Criminal Violations," and DOD Directive 5210.50,

"Unauthorized Disclosure of Classified Information to the Public" and implementing issuances?
5. Are investigations conducted and counterintelligence reports made where necessary in connection with unauthorized absentees?

Access

1. Before access to classified information is granted, is the person determined who will possess the appropriate level security clearance and need-to-know, and is the person given an initial security briefing that includes the requirement to execute a nondisclosure statement?
2. Have procedures been established by the head of each component to prevent unnecessary access to classified information?
3. Is there a demonstrable need for access to classified information before a request for a personnel security clearance is initiated to ensure that the number of people cleared and granted access to classified information is maintained at the minimum number consistent with operational requirements and needs and that it has nothing to do with rank or position?
4. Does the final responsibility for determining whether a person's official duties require possession of or access to any element or item of classified information, or whether the individual has been granted the appropriate security clearance by proper authority, rest upon the individual who has authorized possession, knowledge, or control of the information and not upon the prospective recipient?
5. Unless waived by appropriate authority, is the two-person rule in effect for work in areas where Top Secret information is accessible?
6. Have heads of components designated appropriate officials to determine, before the release of classified information to persons outside the Executive Branch, the propriety of such action in the interest of national security and assurance of the recipient's trustworthiness and need-to-know?

7. Are proper procedures followed with respect to access by historical researchers?
8. Is classified information released to foreign nationals, foreign governments, and international organizations only when authorized under the provisions of the National Disclosure Policy and DOD Directive 5230.11, "Disclosure of Classified Military Information to Foreign Governments and International Organizations"?
9. Is access to COMSEC information by foreign persons and activities in accordance with policy issuances of the National Telecommunications and Information Systems Security Committee?
10. Have procedures been established to control access to classified information by visitors?
11. Do classified visit notifications meet the minimum requirements?

Dissemination

1. Have procedures been established consistent with DOD 5200.1-R for the proper dissemination of classified material?
2. Is particular emphasis placed on traditional need-to-know measures to aid in the strict control of classified information?
3. Has prior approval been obtained for special access requirements with respect to access, distribution, and protection of classified information?
4. Is classified information, which was originated by a non-DOD department or agency, not disseminated outside DOD without the consent of the originator?
5. Is DOD Top Secret information being disseminated outside of the Department with the consent of the originating DOD component, or higher authority?
6. Is Top Secret information, whenever segregated from classified portions bearing lower classifications, being distributed separately?
7. Are dissemination requirements being met with respect to other types of classified information?

8. Are standing distribution requirements for classified information and materials, such as distribution lists, being reviewed at least annually to verify the recipients' need-to-know?
9. Is the handling, use, and assignment of code words, nicknames, and exercise terms in accordance with DOD 5200.1-R?
10. Is the use of classified information in scientific and technical meetings subject to the provisions of DOD Directive 5200.12, "Policy on the Conduct of Meetings Involving Access to Classified Information"?

Accountability and Control

1. Are Top Secret accountability registers maintained by each office originating or receiving Top Secret information, and do the registers reflect all required information?
2. Are Top Secret documents and material numbered serially and marked to indicate individual copy numbers, for example, copy 1 of 2 copies?
3. Does each Top Secret document or item of material have appended to it a Top Secret Disclosure Record?
4. Are all Top Secret documents and material inventoried at least once annually?
5. Is Top Secret information retained only to the extent necessary to satisfy current requirements; do custodians destroy nonrecord copies of Top Secret documents when no longer needed; and are record copies of documents that cannot be destroyed reevaluated and, when appropriate, downgraded, declassified, or retired to designated records centers?
6. Are Top Secret documents and material accounted for with a continuous chain of receipts?
7. Have administrative procedures been established by each DOD component for controlling Secret information and materials?
8. Does the Secret control system provide a means to ensure that Secret material sent outside a major subordinate element (the activity) of the DOD component concerned has been delivered to the intended recipient by use of a receipt or through

hand-to-hand transfer when the receiving party acknowledges responsibility for the Secret material?
9. Does the Secret control system provide a record of receipt and dispatch of Secret material by each major subordinate element?
10. Have administrative controls been established to protect Confidential information?
11. Have procedures been developed to protect incoming mail, bulk shipments, and items delivered by messenger until a determination is made whether classified information is contained therein?
12. Depending on the circumstances, are classified working papers marked and safeguarded properly?
13. Are portions of documents and materials that contain Top Secret information normally reproduced only with the consent of the originator or higher authority?
14. To the extent possible, have DOD components established classified reproduction facilities where only designated personnel can reproduce classified materials and institute key control systems for reproduction areas?
15. When possible, are two people involved in the reproduction process to help assure positive control and the safeguarding of all copies?
16. Is the copying of documents containing classified information minimized?
17. Are officials authorized to approve the reproduction of Top Secret and Secret information designated by position title, and do they review the need for reproduction of classified documents and material with a view toward minimizing reproduction?
18. Has specific reproduction equipment been designated for the reproduction of classified information, and have reproduction rules been posted on or near the designated equipment?
19. Are notices prohibiting reproduction of classified information posted on equipment used only for the reproduction of unclassified information?
20. Have DOD components ensured that equipment used for reproduction of classified information does not leave latent images in the equipment or on other material?

21. Are all copies of classified documents reproduced for any purpose, including those incorporated in a working paper, subject to the same controls prescribed for the original document?

Transmission and Hand-Carrying

1. Is classified information transmitted or transported in accordance with the requirements for each security classification?
2. Does the preparation of classified information for transmission, shipment, or conveyance meet minimum requirements?
3. Is Top Secret information transmitted under a chain of receipts covering each individual who gets custody?
4. Is Secret information covered by a receipt when transmitted to foreign governments/embassies and between major subordinate elements of DOD components and other authorized addressees except during hand-to-hand transfer when the recipient acknowledges responsibility for the Secret information?
5. Is Confidential information covered by a receipt when transmitted to foreign governments/embassies or upon request?
6. Are appropriately cleared personnel, who are authorized to escort or hand-carry classified material, complying with the minimum storage requirements?
7. Are the general restrictions concerning escort or hand-carrying classified material adhered to—including not leaving material, under any circumstances, unattended while it is being carried in a private, public, or government conveyance?
8. Do individuals who are authorized to hand-carry or escort classified material receive an appropriate briefing, and are they required to sign a statement acknowledging receipt of such briefing?
9. Is classified material authorized to be hand-carried aboard a commercial aircraft only when there is neither time nor means available to move the information in the time required to accomplish operational objectives or contract requirements, and other requirements/procedures have been implemented?

10. Is the escorting or hand-carrying of classified information aboard commercial passenger aircraft approved by appropriate authority?

Disposal and Destruction

1. Is documentary record information disposed of or destroyed in accordance with DOD component record management regulations?
2. Is nonrecord classified information destroyed when it is no longer needed, in accordance with proper procedures?
3. Do destruction procedures incorporate a means to verify the destruction of classified information?
4. Are approved methods of destruction used?
5. Have procedures been instituted that ensure that all classified information intended for destruction actually is destroyed beyond reconstruction?
6. Are appropriate destruction procedures followed for each level of classified material destroyed?
7. Are burn bags controlled in a manner designed to minimize the possibility of their unauthorized removal and the unauthorized removal of their classified contents prior to actual destruction?
8. Are records of destruction, which are required for Top Secret information, signed by two cleared persons?
9. When not required, if a record of destruction is used for Secret material, does at least one cleared person sign the record?
10. Do two cleared persons witness the destruction of Secret material when a record of destruction is not used?
11. Is other classified waste—such as handwritten notes, carbon paper, typewriter ribbons, and working papers—destroyed properly when no longer needed?
12. Are classified documents, which are not permanently valuable records, destroyed after five years from the date of origin, unless longer retention is authorized by and in accordance with DOD component record disposition schedules?
13. Does the head of each activity establish at least one cleanout day each year in which a portion of the work performed in

every office with classified information stored is devoted to the destruction of unneeded classified holdings?

Security Education

1. Have security education programs been established to meet basic objectives?
2. At minimum, are all personnel, authorized or expected to be authorized access to classified information, provided an indoctrination on all of the essential principles, practices, and procedures?
3. Are personnel advised of the adverse effects to the national security that could result from unauthorized disclosure and of their personal, moral, and legal responsibility to protect classified information within their knowledge, possession, or control?
4. Are personnel indoctrinated in the principles, criteria, and procedures for the classification, downgrading, declassification, marking, control and accountability, storage, destruction, and transmission of classified information and material, as prescribed in DOD 5200.1-R, and alerted to the strict prohibitions against improper use and abuse of the classification system?
5. Are personnel informed of the techniques employed by foreign intelligence activities in attempting to obtain classified information, and their responsibility to report such attempts?
6. Are personnel advised of the penalties for engaging in espionage activities?
7. Are personnel informed of the penalties for violation or disregard of the provisions of DOD 5200.1-R?
8. Are personnel instructed that individuals having knowledge, possession, or control of classified information must determine, before disseminating such information, that the prospective recipient has been cleared for access by competent authority; needs the information in order to perform his or her official duties; and can properly protect (or store) the information?
9. Are DOD personnel who are granted a security clearance not permitted to have access to classified information until they

have received an initial security briefing and have signed a nondisclosure agreement?
10. Have programs been established to provide, at a minimum, annual security training for personnel having continued access to classified information?
11. Are personnel who have had access to classified information given a foreign travel briefing before travel to alert them to their possible exploitation?
12. Upon termination of employment, administrative withdrawal of security clearance, or contemplated absence from duty or employment for sixty days, are DOD military personnel and civilian employees given a termination briefing, required to return all classified materials, and to execute a Security Termination Statement?
13. Is the nondisclosure agreement briefing pamphlet (DOD 5200.1-PH-1) being used?
14. When an individual refuses to execute a Security Termination Statement, is that fact reported immediately to the security manager of the cognizant organization concerned, and, in any such case, is the individual involved debriefed orally?
15. Is the fact of a refusal to sign a Security Termination Statement reported to the Director, Defense Investigative Service?

Foreign Government Information

1. Are the classification, declassification, marking, and protective requirements for foreign government information being met?
2. Is foreign government information in DOD documents controlled in a way that assures the avoidance of premature declassification?
3. Are foreign government-classified documents that do not have English language classification markings remarked with the U.S. equivalent classification?

Administrative Sanctions

1. Are DOD military and civilian personnel subject to administrative sanctions for knowingly, willfully, or negligently committing security violations?

2. Do commanders and supervisors, in consultation with appropriate legal counsel, use all appropriate criminal, civil, and administrative enforcement remedies against employees who violate the law and security requirements in DOD 5200.1-R and other pertinent DOD issuances?
3. Is appropriate and prompt corrective action taken whenever a knowing, willful, or negligent security violation occurs or repeated administrative discrepancies or repeated disregard of the requirements of DOD 5200.1-R occur?
4. Are knowing, willful, or negligent security violations reported to the appropriate DOD level information security office?
5. Are reports made to appropriate counterintelligence, investigative, and personnel security authorities concerning any employee who is known to have been responsible for repeated security violations over a period of a year, for appropriate evaluation, including readjudication of the employee's security clearance?

Bibliography

Berry, Joseph A. *Interior Alarms.* Presented at ASIS annual convention and seminar. Washington, DC, 1993.

Defense Intelligence Agency. Date unknown. *Information Security Program,* DIA Instruction 240-110-8. Washington, DC: Defense Intelligence Agency.

Department of the Army. Date unknown. *Designing for Security,* TM 5-853-1. Washington, DC: Department of the Army.

Department of the Army. 1979. *Physical Security,* FM 19-30. Washington, DC: Department of the Army.

Department of the Army. 1983. *Information Security Program,* AR 380-5. Washington, DC: Department of the Army.

Department of the Army. 1986. *Risk Analysis for Army Property.* DA Pam 190-51. Washington, DC: Department of the Army.

Department of the Army Corps of Engineers. 1982. *Intrusion Detection Equipment,* U.S. Army Corps of Engineers Guide Specification, Military Construction [CEGS-16750]. Washington, DC: Department of the Army Corps of Engineers.

Department of Commerce. 1974. *Guidelines for ADP Physical Security and Risk Management,* FIPS Pub 31. Washington, DC: Department of Commerce.

Department of Commerce. 1979. *Guideline for ADP Risk Analysis,* FIPS Pub 65. Washington, DC: Department of Commerce.

Department of Defense. Date unknown. *Department of Defense Industrial Security Program,* DoD Directive 5220.22. Washington, DC: Department of Defense.

Department of Defense. Date unknown. *Industrial Security Manual,* DoD Regulation 5220.22-R. Washington, DC: Department of Defense.

Department of Defense. Date unknown. *Industrial Security Manual for Safeguarding Classified Information,* Department of Defense 5220.22-M. Washington, DC: Department of Defense.

Department of Defense. Date unknown. *Inspection Checklist—Department of Defense Information Security Program.* Washington, DC: Office of the Under Secretary of Defense for Policy.

Department of Defense. Date unknown. *Security Requirements for Automatic Data Processing (ADP) Systems,* DoD Directive 5200.28. Washington, DC: Department of Defense.

Department of Defense. Date unknown. *Security Sponsorship and Procedures for Scientific and Technical Meetings Involving Disclosure of Classified Military Information,* DoD Directive 5200.12. Washington, DC: Department of Defense.

Department of Defense. 1978. *Physical Security Equipment—R&D Programs,* 1st Interdepartmental Meeting. Washington, DC: Department of Defense.

Department of Defense. 1984. *Information Security Program,* DoD Regulation 5200.1-R. Washington, DC: Department of Defense.

Department of Defense Security Institute. Undated draft. *Handbook on Intrusion Detection Systems.* Richmond, VA: Department of Defense Security Institute.

Department of the Navy. Date unknown. *Physical Security and Loss Prevention,* OpNav Instruction 5530.14A. Washington, DC: Department of the Navy.

Department of the Navy. Date unknown. *Property Control,* NDW Instruction 5512.5. Washington, DC: Department of the Navy.

Federal Law Enforcement Training Center. 1990. *Advanced Physical Security Training Program,* course 2617.01. Glynco, GA: Federal Law Enforcement Training Center.

Inspection checklists from various sources collected over a decade, from the author's personal reference files.

McCullough, Dennis J. 1993. *Security Surveys,* presented at annual ASIS Seminar. Washington, DC.

Physical Security Controls, 1 page listing, author's personal files.

Roper, C. 1983. *Management Oversight in Developing a Physical Security Inspection Program,* draft paper. Arlington, VA.

Roper, C. 1985–1986. *PSIP Thoughts—H336 Physical Security Inspection Teams,* personal notes, Arlington, VA.

U.S. Army Institute for Professional Development. 1976. *Security Management II,* Army correspondence course, Subcourse MP0078, Edition 6. Newport News, VA:U.S. Army Institute for Professional Development.

U.S. Army Intelligence Center. 1975. *Security Alarm Devices 1,* instructor reference folder. Ft. Huachuca, AZ: U.S. Army Intelligence Center & School.

U.S. Army Military Police School. 1991. *Employment/Evaluating Operations of Intrusion Detection Systems [IDS].* Ft. McCellan, AL: U.S. Army Military Police School.

Index

Access checklist, 271–72
Access control, 54–55, 117–27, 167–81
   computer rooms and facilities, 169–73
   media storage libraries, 169, 170, 173–74
   remote terminals, 174–75
Accountability and control checklist, 273–75
Acoustical systems, 157, 158–59
   See also Intrusion detection systems; sensors
Administrative sanctions checklist, 278–79
Air conditioning checklist, 259
AIS security inspection checklists, 217–39
Alarm systems, 62, 82, 147, 149–50
   See also Intrusion detection systems; sensors
Alarmed areas, classified material and information and, 131–32
Alarms, nuisance, 152, 157, 158, 162
Anti-Compromise Emergency Destruct (ACED) equipment, 198
Area protection, 153–54
   See also Sensors

Asset protection and loss control, 207–210
Assistance visit
   See Inspection
Audit
   See Inspection
Automated Information Systems (AIS), 20, 163, 167–81
   checklist, 241–79
   main computer facilities, 169–73
   media storage libraries, 173–74
   remote terminals, 174–75

Background checks, 120, 123
Backup systems, emergency, 72, 75–76, 151, 162
Badge blanks, 120
Badges
   See Identification badges
Barbed wire, 59–60, 61
Barriers, protective, 53–69, 81–83, 95, 131
   and Automated Information Systems, 169, 171, 179–81
   lighting, 73
   in vault construction, 88

286   Index

Bars, metal, 53, 62, 65–66, 86, 88, 103, 171
Bomb threats, 67, 202–204
  reporting forms, 202
Briefings, inspection, 36–41
Building perimeters checklist, 232–34
Buildings, 61–62, 75
  See also Facilities

Ceilings, 90–96, 160
Chain-link fences, 57–60, 172
Checklists, 217–279
  See also individual checklists
Checkpoints, 54, 83, 123
Chief of physical security, 7, 20–21
Chief of security, 7, 20
Classification challenges and resolutions checklist, 265
Classification guides checklist, 263–65
Classification in industrial operations checklist, 265–66
Classification policy and principles checklist, 261–63
Classified documents checklist, 266–68
Classified information
  See under Information and materials, types of
Cleaning crews, 123, 127, 208, 209
Clearances, 120, 123
Closed-circuit television (CCTV), 3, 163–65, 172, 174, 199
Combination locks, 102–103, 181
  and key control, 107–108
  and security containers, 132–34, 139–41, 145–46
  standards, 90, 132, 139–41
Combination padlocks
  See under Padlocks
Communications security (COMSEC), 196
Compromise of classified information checklist, 270–71
Computer diskettes, security of, 144
Computers
  location of central complex, 169–70
  remote terminal security, 174–75
  See also Automated Information Systems
Concertina wire, 59–60, 61
Confidential information
  See under Information and materials, types of

Construction standards and requirements
  facilities, 85–87, 94–95
  strong rooms, 92–93
  vaults, 87–92, 93–94
Contract guard forces, 184, 185, *187*
Controlled areas
  See Security areas, types of
Counterforce testing
  See Vulnerability testing
Countersabotage checklist, 248
Crossbars, 63, 64, 66
Custodial responsibilities, 143–46
  checklist, 269–70
Customer Engineers (CEs), 172–73
Cypher locks, 102–103, 181

Data collecting and reviewing, *10*, 13–17, 31–34, 41–42
Deadbolt locks, 101, 104
Declassification and downgrading checklist, 266
Department of Defense (DOD), 81, 130, 184
  checklist, 260–79
Disposal and destruction of information checklist, 276–77
Dissemination checklist, 272–73
Document accountability systems checklist, 247
Document and information security checklist, 245–46
Doors
  access control, 171, 173, 174, 179
  exterior, 86, 160
  to fire escapes, 87
  hardware, 99, 113–14
  locking devices, 98–99, 100, 103
  security and security upgrades, 62, 95, 96
  standards, 112
  strong rooms, 93
  types and materials, 99, 113
  vaults, 87–88, 90, 91, 94
Ducts, 69, 88–89, 93, 171

Electrical power, 75–76
Electrical power checklist, 259
Electronic access control system (EAS), 122

Emergency backup systems
   and intrusion detection systems, 151, 162
   lighting, 72, 75–76
Entranceways, lighting, 74
Environmental security protection, 175
Essential Elements of Information, 8, 27, *28*
Exclusion area
   See Security areas, types of
Exit briefings, inspection, 36, 39–41

Facilities
   admittance requirements, 123–24
   construction standards and requirements, 85–87, 94–96
   inspection, 9–11, 15–17
   management, 16, 18–20, 207
   personnel, 17–18, 123, 208
   security, 54–55, 199–206, 211–15
   threats to, 199–206
   See also Vehicular access and controls; identification badges
Facility entries, miscellaneous, 69, 86–88, 93, 95, 171, 210
   crawl spaces, 171, 210
   ducts, 88, 93, 171
   fire escapes, 87
   stairwells, 86, 87
   vents, 171, 210
False alarms, 152, 157, 158
Fences, 54–60, 73–74, 83, 172, 157–58
   See also Barriers, protective
Field program management, 21–22
Fire escapes, 87
Fire exposure and risk checklist, 256–58
Fire, prevention and suppression, 88, 94, 171, 173, 199, 200
Floodlights, 73, 74
Floors, 89–91
Foreign government checklist, 278
Freedom of Information Act (FOIA), 32

Gates, 74, 164
General Purpose Barked Tape Obstacle (GPBTO), 60
   See also Concertina wire
General Services Administration (GSA), 195–96

security containers, 130–32, 135, 137–39
Glare Projection Method, lighting, 73
Glass
   See Windows
Government classified material and information, 129–34
Grillwork, 62–65, 66, 171
Group 1-R combination locks, 102, 140
Guard posts, 53–55, 73, 74, 111, 189
Guards
   See Security guard forces

Hardware
   barriers, protective, 64–65
   construction standards, 88, 94–95
   doors, 114
   lock and key, 98–100, 132
Hardware security checklist, 249
Hinge assemblies, 99, 113–14

Identification and badges checklist, 222–25
Identification badges, 54, 74, 82–83, 119–27
Information and materials, control of access, 79, 82, 129–31, 142–43
   custodial responsibilities, 143–44
   during working hours, 144–45
   evacuation and destruction, 144, 196–98
   facility data, 19
   government classified information, 79, 129–34
   See also Security containers and storage areas
Information and materials, types of
   classified, 82, *89*, 130–34, 137, 141–45, 173, 196
   confidential, 131, 135, *142*, 198
   corporate sensitive, 19, 129, 196, 197
   government classified, 79, 129, 196
   low-level, 131
   secret, *89*, 132, 135, *142*, 198
   sensitive, 129, 143, 144, 145, 173, 198
   sensitive unclassified, 79
   top secret, *89*, 131, 132, *142*, 198
   unclassified, 19, 79, 134, 136, 143, 144

Inspection, 23–51
  briefings, 36–41
  data collection and analysis, 31–34,
    41–42
  planning process, 11–22
  See also Preinspection
Inspection report,
  See Report of Inspection
Inspection team
  administrative preparations, 26–27
  and automated information systems,
    167–68
  facility security considerations, 38–39
  frequency of meeting during inspection, 34
  organization and duties, 7–8, 17,
    20–22, 80–81, 85
  training and experience, 3–6
  working with facility personnel,
    10–11, 17–18, 29, 80
Internal Security Act of 1950, 81
Intrusion detection systems (IDS),
    147–62, 173, 180, 199
  checklist, 225–26
  See also Alarm systems; sensors

Janitorial workers
  See Cleaning crews
Joint Service Interior Intrusion Detection
  Systems (JSIIDS), 161–62
Junction boxes, 152

Key control, 19, 80–81, 99–100, 107–12,
    133, 180–81, 209–10,
  checklist, 227–28
Key cores
  See Lock cores and cylinders
Key hardware
  See Lock and key hardware
Keyboard control locks, 174

Latchbolts, 104
Leafs
  See Hinge assemblies
Libraries, media storage, 169,
    173–74
Lighting, protective, 54, 71–77, 180
  checklist, 228–30
Limited areas
  See Security areas, types of

Lock and key control, 180–81
Lock combinations
  security, 108, 145–46
  selecting and changing numbers, 107,
    145–46
  See also Padlocks; combination locks
Lock cores and cylinders, 98–102, 105,
    133, 181
Lock parts, basic, 103–104
Lock strikes, 98, 103–107
Locking bar containers, 131, 134
  See also Security containers and storage areas
Locking devices, 97–115
  considerations for, 98–101
  defeating, 97, 99, 105
  for doors, 93–103, 106, 112–14
  electrically actuated, 146
  high-security, 100, 210
  keyed locks, 104–105, 108
  levels of security, 98, 112
  lock parts, basic, 103–104
  locking bars, 132
  locking bolts, 103–104
  for security containers, 130–33
  types of, 101–103
  for vaults and strong rooms,
    90–91, 93
  for windows, 115
  See also Combination locks; padlocks;
    lock strikes
Lock picking, 99, 105
Locks, rotation of, 98, 133
Loss control, 207–10

Map and plan file containers
  See Security containers and storage
    areas
Marking and handling checklist, 246
Marking checklist, 266–68
Master keys, 99, 105, 109–10, 112,
    133, 209
Medeco high-security keyed
  cylinders, 103
Media destruction and declassification
  checklist, 247
Metal bars, 53, 62, 65–66, 86, 88,
    103, 171
Media storage libraries, 169, 173–74
Military police, 184

Mortise locks, 101, 112
Motion detection devices, 156–57, 158–59
   See also Intrusion detection systems; sensors

Narrow-stile locks, 101
Natural disasters, 175
   checklist, 259–60
National Fire Protection Association (NFPA), 88
National Industrial Security Program Operating Manual (NISPOM), *xi*, 29
Nonrestricted areas, 81, 83
   See also Security areas, types of
Nuisance alarms, 152, 157, 158, 162

Package control, 19, 80–81
Padlocks
   combination, 93, 103, 107–108, 131, 133, 140–41
   high-security, 103, 112, 134
   keyed, 103, 107, 133
   rotation of, 107
   Sargent and Greenleaf models, 103, 131, 140–41
Panic bars, 87
Parking lots and permits, 74, 124–27
   See also Vehicular access and control
Passkeys, 209
Penetration testing
   See Vulnerability testing
Perimeter protection, 154
   See also Sensors
Perimeter security checklist, 230–32
Perimeter testing
   See Vulnerability testing
Personal computer diskettes, security of, 144
Personal property, asset protection and loss control, 208–209
Personnel
   access control, 119–22
   nonsecurity, 10–11, 17–18, 109, 119–23, 208, 210
   security, 17–18, 53–56, 72–73, 147, 152, 200, 207, 210
   See also security guard forces
Personnel security checklist, 244–45
Photoelectric detectors, 155, 158–59

See also Intrusion detection systems; sensors
Physical security checklist, 254–56
Physical security chief, 7, 20,
Physical Security Inspection Program (PSIP)
   data collecting and analysis, 30–34, 41–42
   facility criticality and vulnerability, 211–15
   field program management, 1–22
   overview, 1–51
   Report of Inspection, 43–50
   team organization and duties, 7–8, 20–22, 80–81
Picture identification card
   See Identification badges
Pipes and fuel lines, 76–77, 171
Point of contact (POC) inspection, 33–34
Point protection, 149, 151, 153–54
   See also Sensors
Preinspection
   briefings, 37–38
   data collection, 31–32
   planning and preparation, 25–28
   sample time frame, *30*
   scheduling, 27–31
Printer ribbons, security of, 144
Program management checklist, 260–61
Proprietary guard forces, 184, 185, 186, *188*
Protection, man-hours and man-minutes, 138–40
Provost Marshall Office/Military Police (PMO/MP), 29
Proximity systems and detectors, 156, 157–59
   See also Intrusion detection systems; sensors
Public displays, protection of, 208

Relocking devices
   See Locking devices
Report of Inspection (ROI), 13, 40–50
Restricted areas, 79–83
   See also Security areas, types of
Resurveys, 26
Rim deadlocks, 112
Roof access, 67–69, 86
Room protection, 149, 153–54
   See also Sensors

Sabotage and espionage, 200–201
Safety glass
　See Windows
Sargent and Greenleaf, 103, 131, *136*, 140–41
Secret information
　See under Information and materials, types of
Security areas, types of, 79–83, 176–78
Security containers and storage areas, 129–46, 210
　government classified information, 131–34
　GSA standards, 130–32, 137–38
　locking bar containers, 131, 134
　media storage libraries, 169, 173–74
　supplies and equipment, 210
　vaults and rooms, 129, 141–43, 145
　See also Vaults
Security control and personnel checklist, 246–47
Security duties checklist, 253–54
Security education checklist, 277–78
Security guard forces, 15, 72, 73, 132, 161, 183–93
　checklist, 234–36
　and protective barriers, 53, 54, 57
　See also under Personnel
Security hours and work hours, 123–24, 144
Security, levels of, 98, 100, 102, 108, 112
Security office
　checklist, 217–20
　concerns of, 18–20, 168, 190–91
Security patrols, 180
Security personnel
　See under Personnel
Security team
　See Inspection team
Security upgrades, 94–96
　See also Construction standards and requirements
Sensitive information
　See under Information and materials, types of
Sensors, 149–61
　electromechanical, 154–55
　electronic, 155–58
　and roof openings, 86
　types of protection, 149, 151, 153–54

vaults, 96, 160–61
　See also Intrusion detection systems; alarm systems
Site security managers, 22–23
Site security personnel, 23
Site waivers, 38
Skeleton keys, 105
Software security checklist, 249–50
Space protection, 153–54
　See also Sensors
Spot protection, 149, 151, 153–54
　See also Sensors
Sprinkler systems, vaults, 88
Stairwells, 86, 87
Standard operating procedure (SOP), 30
Steel filing cabinets, 132, 134
Storage and storage equipment checklist, 268–69
Storage areas
　See Security containers and storage areas
Strikes and strike plates
　See Lock strikes
Strong rooms, 92–93
System certification and testing checklist, 243–44
System design checklist, 248
System security procedure checklist, 242–43

Tape library checklist, 250–53
Technical surveillance countermeasures checklist, 247–48
Threat data, 31, 40, 46
Threats, to facilities, 199–206
Top secret information
　See under Information and materials, types of
Transmission and hand-carrying checklist, 275–76
Tumbler wheels, 132–33, 138–40
Typewriter ribbons, security of, 144

Unclassified information
　See under Information and materials, types of
Underwriter's Laboratories (UL), 90, 139–41, 150
Uninterruptable power service (UPS), 176

Utilities, 76–77, 175–76
  checklist, 236–38

Vaults, 129–33, 137–39, 141–43
  alarm systems, 95–96
  construction standards and requirements, 87–96
  detection devices, 161–62
  escape devices, 88, 92
  See also Security containers and storage areas
Vehicles
  access and control, 74, 124–27
  checklist, 238–39
  garages, 112
Vibration detection, 155
  See also Intrusion detection systems; sensors
Visibility
  See Lighting, protective
Visitor control, 172, 208
Vulnerability testing, 34–36

Walls
  as protective barriers, 60–61

  sensor applications, 160
  in strong rooms, 92
  in vaults, 87, 89–90, 91
Warded locks, 105
Warning signs, 180
Water damage exposure checklist, 258
Water pipes, and media storage libraries, 173–74
Windows
  foil and tape, 154
  plastic used in, 66, 67
  in protective barriers, 62–67
  protective glass for, 66–67
  in roofs and skylights, 65–66
  sensors, 66, 154, 160
  not in vaults and strong rooms, 93–94
Wire, barbed or concertina, 59–60, 61
Wirelines, protection against tampering, 152–53
Work hours
  See Security hours and work hours
Workmen, and asset protection and loss control, 210